Philosophy of Science

Perspectives from Scientists

Philosophy of Science

Perspectives from Scientists

Paul Song

University of Massachusetts Lowell, USA

World Scientific

NEW JERSEY · LONDON · SINGAPORE · BEIJING · SHANGHAI · HONG KONG · TAIPEI · CHENNAI · TOKYO

Published by

World Scientific Publishing Co. Pte. Ltd.

5 Toh Tuck Link, Singapore 596224

USA office: 27 Warren Street, Suite 401-402, Hackensack, NJ 07601

UK office: 57 Shelton Street, Covent Garden, London WC2H 9HE

Library of Congress Control Number: 2022942199

British Library Cataloguing-in-Publication Data
A catalogue record for this book is available from the British Library.

PHILOSOPHY OF SCIENCE
Perspectives from Scientists

ISBN 978-981-126-116-9 (hardcover)
ISBN 978-981-126-117-6 (ebook for institutions)
ISBN 978-981-126-118-3 (ebook for individuals)

For any available supplementary material, please visit
https://www.worldscientific.com/worldscibooks/10.1142/12990#t=suppl

Desk Editor: Nur Syarfeena Binte Mohd Fauzi

Typeset by Stallion Press
Email: enquiries@stallionpress.com

To my wife Peng Guo and daughter Lorna

Contents

List of Figures

Chapter 1

Introduction

1.1 Different Perspectives in Philosophy of Science

Just as there are philosophers who love science, there are scientists who love philosophy. However, their contributions to the field of philosophy of science are unequal. Over the past hundred years, the central questions and mainstream theories of philosophy of science have been framed by philosophers, linguists, and historians with little input from scientists. One exception is Thomas Kuhn, who, although not a scientist, had rigorous entry-level training in science and made significant contributions to the field of philosophy of science. On the other hand, there are several scientists who have contributed to some peripheral or specialized subtopics in philosophy of science. By studying the conventional theories of philosophy of science and the scientific method, it is clear to me that philosophers and scientists have very different perspectives on philosophy of science. Some of these differences are due to differences in training and experiences, but other differences are more fundamental in philosophy! The difference starts from fundamental questions of what science is and what scientists do. On the one hand, to a philosopher, "science is just the attempt to understand, explain, and predict the world we live in" using "particular methods that scientists use to investigate the world" (Okasha, 2016). Accordingly, the objective of philosophy of science is to find these "particular methods", which they call "scientific methods", for

1

scientists to follow in their scientific pursuits. To scientists, on the philosophical level, science is a process of developing new knowledge, and scientists are to invent new knowledge while filtering out false knowledge. The difficulty in this process is that the newly invented knowledge could be wrong, and the knowledge filtered out could be correct! One of the critical questions to a scientist is the following: How can we prevent these mistakes from happening?

Most existing theories of philosophy of science are natural extensions of epistemology in philosophy. However, epistemology concerns knowledge in general. Science, instead, is mostly about inventing and developing new knowledge and less about the nature and scope of knowledge that is defined by their profession. The questions for philosophy of science should be a set of focused ones in epistemology. Over centuries, scientists have developed, but not well theorized, a system to handle special situations in science. This system defies the standard theories of epistemology and, unfortunately, is mostly unknown to philosophers of science. This book aims to theorize and present this system, i.e., a new theory of science in philosophy of science, although not yet a complete theory of philosophy of science.

Many people think that science, in a nutshell, inquires into the "root of the matter". While this may be true and sounds very inspiring to curious minds, it is only a small part of what scientists do. Philosophers ask deep questions, some of which can be so deep that no one can answer them. In addition to asking questions, scientists have to find answers to practical questions. To answer a question, an immediate new question is whether the answer is correct or not.

The conventional idea of philosophy of science tells us that scientists need to be rational by proving an idea with evidence. Theories of philosophy of science cite many historical stories centuries after the events like commentators that tell the general public who did something right or something wrong for precisely what reasons. They explain why science has successfully invented new knowledge that other investigative disciplines should emulate. According to these theories, if a story of science development is not rational, we have got it wrong. Conventional theories of philosophy of science offer their help to make a "rational reconstruction" of the *history* of science in order to ensure our theories, information, logic, and critical reasoning are correct. To most people, especially science enthusiasts, rationality and evidence should lead us to the truth, although some may have reservations to "reconstruct history".

On the other hand, scientists invent new knowledge and ideas. To them, it is often not so clear what is right or wrong at the time of an event. This is because they predominantly deal with other scientists who are also investigating the same problem. At the philosophical level, if all scientists are rational and prove their ideas with justified evidence, one may think they should all come to the same answer or solution to a problem, just like finding the same solution to an equation in mathematics. But then, one finds that scientists have different ideas or answers and debate among themselves. If they debate heatedly, some rationalizations must be wrong or evidence incomplete or ambiguous. Therefore, an essential task for a scientist is to find out which rationalization and evidence are *more likely* to be correct. What makes it so challenging to figure out the correctness of a rationalization? Shouldn't evidence be an indisputable hard fact, a curious mind may ask? In reality, unfortunately, the evidence is rarely a hard fact itself and often depends on the *interpretation* of information or facts based on a *plausible* rationalization or idea. It is often that different rationalizations/ideas interpret the same information or observation as evidence that supports opposite theories!

The history of science has shown that science can eventually pick the right idea and rationalize it accordingly. But this is usually done long after the dust has settled; however, each of us has only a blinkingly short life in the long history of civilization. Each scientist has to decide when a question is still being debated versus learning about accepted information and knowledge from, e.g., a textbook. No scientists want to waste their talents by working on a wrong idea. Everyone wants to make a correct decision to either invent new ideas that will positively contribute to and benefit our society or follow and further develop a new idea that is later proven right. This decision is difficult because, at the time of an event, our knowledge and information are incomplete or partial, in contrast to the 20/20 vision viewed from a rearview mirror by historians and philosophers in philosophy of science. The questions and problems that scientists have to face lead them to different perspectives from that of historians and philosophers, as I will discuss in this book.

This book is written for scientists, students, and teachers of science and engineering, and readers who are interested in philosophy and philosophy of science. In this book, I outline a new theory of science that is fundamentally different from the existing theories. The new theory is built upon over three decades of my experience in deliberating the fundamental questions in space physics. I started developing my understanding of the system of science and my theory of science 36 years ago during my graduate program under the guidance of Dr. Christopher T. Russell. He provided many insights that formed the core of my working knowledge of philosophy of science and my understanding of the structure of science.

In 1994, Dr. Eugene Parker called for a *paradigm change* in space physics, which will be described in Section 6.5. I have been very fortunate to work with Dr. Vytenis Vasyliūnas since 2001 on this paradigm change project. While the project involves much physics and mathematics, I found that the two of us have spent much more time discussing questions such as *how to get it right* and *how to know we are not wrong*, the central questions of philosophy of science discussed in this book. For scientists, these questions become essential when they can competently handle all technical aspects of an issue and when the research aims at challenging existing paradigms or theories. The experience in this process has strongly influenced the theory developed in this book.

There are significant differences in the interests and concerns between the targeted audience and philosophers. First, philosophers require their terminology to be linguistically consistent with all aspects of their meanings used in histories and dictionaries. In science, a term means what it means in common sense and is defined in the context of a publication. Our discussion is irrelevant to politics, religion, and metaphysics, where ambiguities in interpreting some terms may occur. Some examples of social sciences are used for demonstrative purposes of a theory, but in scientific discussions, political correctness is not a measure to judge the theory's correctness.

Second, the interpretation of many historical events, e.g., the event around Copernicus's theory, discussed in Section 1.4, could be quite different by a scientist. As you will find out by reading this book, scientists allow many other possibilities if they are self-consistent and consistent with significant undebatable facts and any applicable governing scientific principles and laws. Many possibilities presented in this book may not necessarily be consistent with the popular interpretation of a fact discussed in the conventional theories of philosophy of science. As stated above, evidence often depends on interpreting an observation based on a particular rationalization or theory. Taking an *interpretation* of a fact, i.e., *evidence*, as the *fact* itself is often a starting point of bias or flaw in science or other types of investigation, such as philosophy of science.

Third, in conventional philosophy of science, a theory is often required to be general and universal primarily because of the guiding idea required by *deductive* logic, which will be discussed in Section 4.2 and Chapters 7 and 9. In this case, any counterexample is fatal to the theory. However, science comprises many disciplines of various natures and states of development. Some are more quantitative where possible interpretations of a fact are greatly limited by a quantitative comparison of the prediction with observation. However, many disciplines are still mostly descriptive, where the fact or observation may still be subject to multiple interpretations. In principle, it is easier to propose a qualitative theory or model, but the uncertainty and ambiguity of the theory or model can also be huge. One can easily find either supporting examples or counterexamples for a descriptive theory. Even if a theory can explain major observations and facts with evidence that is based on an *interpretation* of the observations, one may be able to find something in detail to argue against the theory based on a *different* interpretation. It is impossible to find a universally true theory of science in philosophy based solely on deductive logic and without allowing exceptions. We hence focus on major observations and allow exceptions to exist, in principle, when we develop the theory of science. As we will see, this notion is against Karl Popper's idea of falsification in philosophy of science.

Because the objective of this book is to *improve* our understanding of science, it is more important to understand why a theory, especially a widely accepted one, does not work than why it works. Therefore, one will find in this book more negative comments than positive comments concerning a theory. One may feel uncomfortable, but this is the essential spirit of science: scientists spend much more time in falsifying an idea, mostly one's own idea.

1.2   Starting Questions

Often a scientist starts an investigation with probing or thought-provoking questions, but many conventional theories of philosophy of science tell us that science begins with a hypothesis or conjecture. Therefore, we differ from the beginning. So, let us start with a few questions as a scientist does and see where these questions lead us.

How do we know that the Earth is not flat but a sphere?

The responses can be as simple as the following: I learned it when I was a child; I have a globe of the Earth; my teacher told me; I saw a photo of the Earth taken from space…

If you respond in this way, you have understood this as a question of *knowledge* — things many people believe true, a concept of philosophy of science to be more carefully defined and discussed in Chapter 3 — or specifically referred to in philosophy of science as *knowledge-that* — literally, something many people know about and believe to be true. In philosophy of science, however, one may question whether what they have been told or have learned is true or not. The answer to this question with evidence and reasoning is referred to as *knowledge-understanding* — literally, things many people not only know about but also understand — in philosophy of science. Obviously, in our everyday life experience, the Earth appears to be flat! One may argue that they were once on a beach (or a cruise ship) and saw a curved ocean horizon across the viewing direction with a distant ship emerging into view. This demonstrates that the ocean surface is curved both in and across the viewing direction. Indeed, this can be considered a very convincing piece of evidence. However, for those who have lived inland, for example, as in a mountainous area and have never seen an ocean, the horizon is rugged. Shouldn't they conclude the Earth to be mostly flat with some mountains and valleys which may be filled with water to become lakes and rivers? At least the water surface of a lake seems unambiguously flat. Doesn't it?

On the other hand, to be a scientist, you are interested in finding new knowledge. Anything that is already known is *not* interesting unless it is used to discover or invent something else new, or you find that the knowledge itself is potentially flawed. In this regard, you should imagine yourself in a time before people knew that the Earth is a sphere or when two groups of people were newly debating about whether the Earth was flat or spherical. With the information and knowledge available at that time, which side would you take in this debate and how could you prove that the Earth is a sphere? If you were a strong proponent of a flat Earth and staked your name on trying to prove that, you would have no problem guessing how people would talk about your findings now. Ancient people, based on the very limited ability to observe with scientific reasoning, actually proved that the Earth is a sphere around the 5th century B.C. How did they do this? I will discuss this in Chapter 4.

What is knowledge?

After considering the last question, you should have noted that "knowledge" is something very useful. If someone proved beyond a reasonable doubt that the Earth is spherical, you do not need to repeat the process of showing evidence and reasoning for it — it is now common knowledge that the Earth is spherical. But there is a problem: Could knowledge be wrong? I recall that people believed that the Earth was flat before someone convincingly showed them that it was a sphere. Furthermore, we have proven that many things previously considered common knowledge are wrong. How do we know if today's common knowledge is not flawed, or if it will hold for a long time, such as our lifetimes, and extend into future generations? I will discuss this problem more in Chapter 3.

What is science? How do we know that science, as people often assume, will bring us correct answers?

The term "science" is derived from the Latin word *Scientia*, meaning the result of logical demonstration, i.e., something that reveals a general and necessary *truth*. This definition only touches a small part of today's real science. Science is *a process* to *invent new knowledge and filter out false knowledge*, as will be discussed in this book. There are tremendous confusing and widespread misconceptions about science in the general public, in addition to the philosophy of science community. Often in public discussions, such as reports, speeches, and debates, if something is mentioned as scientific, it is taken as the *truth* without the need for further discussion or debate, although there is also an undercurrent against science which thinks "scientific" to mean not human. Why is science so powerful, and should it be this powerful? Can a scientific result be wrong? As discussed earlier, many people consider science powerful because it is rational and based on evidence. This reasoning seems very convincing until we realize that there are different ideas and rationalizations. Furthermore, in science, evidence is not always a fact or direct observation; instead, it is an interpretation of that fact or observation, as discussed in Section 1.1. For example, a murder weapon with fingerprints of the murderer on it may seem to be clear evidence that no one can argue about or deny. But a simple fact or observation may be interpreted in multiple possible ways, each of which could be based on a particular theory or rationalization.[1] Which interpretation is correct or more reliable? These questions will be discussed in depth throughout this book, especially in Chapter 9.

Is there a scientific method?

"Of course", most people would answer, "Yes!" because scientific methods have been taught in high school or even earlier. Many of us could elaborate on several methods, such as the famous hypothetico-deductive (H-D) model. With the H-D model, one starts with a *hypothesis* and collects evidence to *deductively* prove that the evidence supports the hypothesis. Done. But as stated at the beginning of this section, scientists start with questions and not hypotheses. Newton did not begin his study with a hypothesis of gravity that needed to be proven, but instead, he questioned, as the legend goes, "Why do apples fall towards the Earth?".

In this book, I will question the very notion that there is a so-called scientific method in science. If you follow the popular scientific method rigorously, unfortunately, you likely reach a wrong conclusion or find

---

[1] In the O. J. Simpson murder case, there was a famous golf glove that was found behind OJ's house and seemed to be his size. OJ was known to be a good golfer. The glove was once soaked with blood but was dry when presented in the courtroom. The DNA samples from the glove were consistent with that of OJ and the murder victims. Could it be used as evidence for OJ being the murderer? In the most dramatic moment of the whole trial, the prosecutor asked OJ to try it on. OJ did, but it did not fit. What conclusion can you draw?

yourselves on the wrong side of a debate, as discussed in an example in Section 1.4. Different methods and reasoning, that may or may not be scientific, will be discussed in Chapter 7. In contrast, scientists derive and present conclusions following scientific reasoning, which is the language that all scientists speak to communicate with each other, both consciously and subconsciously. Although they may disagree on specific scientific points within a debate, they all follow the same set of scientific reasoning. If you want to be a scientist, it is essential to learn about scientific reasoning, which will be discussed in Chapter 9.

How do I get it right? How do I know that I am not wrong?

These two questions are closely related, but the second one is much more difficult to answer. According to the ancient definition of science, a "logical demonstration" can get things right. Therefore, how to "get it right" — how to justifiably prove that something is correct — has been a central question for epistemologists and philosophers of science. However, if everyone knows how to get it right, they should derive the same result, shouldn't they? They should, *IF* they have complete information. But in science, the information is most likely incomplete; there may be many different possible logical demonstrations for the same reason as to why there are multiple sets of solutions when the number of unknowns is greater than the number of equations. How do I decide which logical demonstration is right? Or maybe worse, the problem may involve something too complicated to sort out (perhaps, I cannot pinpoint how many unknowns and how many equations) or some mathematics that is too complex to solve. How can I get it right? I will discuss the problem more extensively in Chapter 9.

If I am lucky enough to have obtained a result with a method or reasoning I think is correct, can I be wrong? How do I know I am not? In other words, how do I know the method and (or) reasoning I am using are correct and flawless? If my answer to these questions is that I am using exactly the same method everyone else is using, can I still be wrong? This problem is undeniable when opposite results are recognized and presented by two scientists on a subject. Shouldn't one of the two be wrong? Or, are both of them right, or both wrong? How do I know that I am *not* the person with the wrong result? These questions are crucially vital for one's scientific career! How can one be so confident and, at the same time, overcome personal biases? A good scientist cannot make a major mistake, or their scientific credibility would be seriously compromised. How to avoid the latter from happening will be discussed in Chapter 11.

What is a Ph.D?

A Ph.D. stands for "Doctor of Philosophy" and is a desirable degree for many academics and professionals, you may respond. But why ask? This is a curious question, you may wonder. There are many different types of Doctoral degrees in addition to a Ph.D., such as a Medical Doctorate (M.D.) or a Doctor of Engineering (Eng. D.). What is unique about a Doctor of Philosophy, and why is it more challenging to get? Why is philosophy so important?

Historically, Ph.D. degrees were designed for those who wanted to become teachers (in churches). But why are teachers so special, needing this particular degree program? What separates them from other doctoral degrees? When answering this last question, many recognize that a teacher is knowledgeable about the subject and can teach it. Teachers must have well-organized thoughts when preparing a curriculum and leading others; they are good at and do not fear public speaking, and can understandably explain complex concepts with simple, clear, and concise language, you may add. Don't forget that teachers are the ones who give you homework assignments and exams! "So what? This is a privilege and not an ability", you may respond. But teachers have the answers to all the questions in homework and exams! "Then so what? They are supposed to know", you may still disagree. What if the teacher gave a wrong answer causing you to fail your test? Wouldn't you challenge the teacher? To avoid this, the teacher has to know how to get the answers right and know why the answers are not wrong, the two questions in the last topic.

After graduation, you no longer have a teacher to tell you what answer is correct and what is wrong. You are on your own to figure out right or wrong. What sets a Doctor of Philosophy apart, no matter their area of expertise, is their ability to find the right answers and make correct decisions. Of course, one can gain this ability through practice and not necessarily through a Ph.D. program. Even if you choose to be a leader within a business or industrial field, rather than in academia, your subordinates will look to you for guidance. Are you able to lead them to success? I will discuss this problem more throughout the book.

In the following three sections, I discuss three examples that are repeatedly used in debates in conventional theories of philosophy of science.

1.3 Raven Paradox

First proposed by Carl Hempel in 1945, this is one of the most famous paradoxes in philosophy of science. As it is described below, you may disagree with the logic, but the purpose, for now, is about what causes the paradox rather than how to resolve it.

Let us assume that we have observed many black ravens, and no raven of other colors has been found. According to our typical way of thinking, we guess that all ravens are black. This must be true because this is what we have *observed* — we observed without wearing sunglasses during sunny days, and we checked that the black color was not painted on. With this information, we may then make a (scientific) hypothesis that "all ravens are black". The scientific method tells us we should start our scientific pursuit to prove it. We could simply find more black ravens and reduce the chance of the hypothesis being wrong. However, we could "deductively" justify this hypothesis according to the hypothetico-deductive method we learned, where the deduction is a formal logic to be discussed in Section 4.2. We reason that because all ravens are black, a non-black thing is non-raven. We then go out looking for supporting evidence for our hypothesis, and when doing so, we find a white swan. Because this swan is non-black and non-raven, our hypothesis holds well. We may also find many other things around us, such as a red coat and a blue pan, that are non-black and non-raven. These items can all be used as evidence to support our hypothesis that "all ravens are black". Therefore, we declare that we have found a large amount of supporting evidence for our hypothesis and that our hypothesis is justified. Because we are good scientists, we go further by searching for all black things and finding ravens among them. This confirms our hypothesis.

So far, every step of reasoning is rigorous deductive logic. We are now ready to announce to the world that our hypothesis of "all ravens are black" is justified or proved. But one thing happens: There are non-black ravens that have been found (due to albinism). Now the question is as follows: Where is the flaw in our proof? We have followed everything we were told about science as faithfully as possible, yet our hypothesis can be disputed, or *falsified*. Hence, this example is a paradox because its hypothesis is both justifiable and disputable. It shows that the H-D scientific method is fallible in this case. You may go to the internet to find various modifications of this method that are supposed to avoid the fallacy. But aren't they fallible too? A fallacy such as this is not allowed in science. If you start questioning the scientific method you have learned, how do scientists avoid the problem? Is the problem specific to this particular case, or is the method fundamentally flawed? We will address this question and discuss the paradox of the ravens in more detail in Chapters 4 and 7.

1.4 Geocentric Versus Heliocentric Theory of Planetary Motions

The paradox of ravens may be a demonstrative example by philosophers of the usefulness of philosophy of science. Copernicus's heliocentric theory may be used to illustrate the difficulty scientists face every day in their research. This is one of the most famous stories widely known by the general public and discussed in philosophy of science; you may have heard about it before. However, the version you heard about has likely been distorted based on conventional theories of philosophy of science. Because it is critically important to understand philosophy of science from the perspectives of scientists, I will give a detailed description of this example in this section. It will be discussed and cited frequently in later chapters. You may find it necessary to reread it as needed.

According to the geocentric theory of the motion of heavenly bodies, the Earth is at the center of the universe, and all celestial bodies circle the Earth. This may be intuitively correct since one sees sunrise and sunset and moonrise and moonset every day. This theory is believed to be developed by Claudius Ptolemy (100–170 AD), an astronomer in Roman Egypt. With various modifications and improvements, this theory held for an extended period without serious challenges until 1543 AD when

Nicolaus Copernicus proposed a heliocentric theory in which all planets, including the Earth, circle the Sun. In philosophy of science, this event marks the start of a scientific revolution because the heliocentric theory is more scientific and has surpassed the geocentric theory. This overall assessment and reasoning, unfortunately, is incorrect or at least question-able to a scientist.

Let us first ignore the question of which theory is right or wrong. When discussing a historical event in philosophy of science, as a scientist, we have to imagine ourselves living during the time of the event and not knowing current beliefs. This is because for the scientific investigations and debates we are conducting, we may not know in our lifetime which theory will eventually be right centuries later. Many popular theories of philosophy of science have been developed based on the knowledge of the future evolution of science, which was unknown at the time of the event. If one follows the theories and uses the evidence available at the time, they will only find themselves to be wrong centuries later. Being scientists at Copernicus's time, without the luxury of knowing the future, we each have to make our own judgment on whether the geocentric theory is (more possible to be) right or the heliocentric theory is. We all want to make correct judgments so that our contributions can be recognized in future research or history. If I make a wrong judgment, on the other hand, espe-cially if the judgment is proven wrong before my retirement, my scientific career and credibility would be fundamentally affected. I can use the best, no matter how incomplete, knowledge and information I have at the time. But the best knowledge and information at the time could be inaccurate or even incorrect. This is the fundamentally different perspective a scientist has from a philosopher who has often based their theories on the informa-tion available centuries later.

Now, we examine the event around Copernicus's theory using the scientific method. We have two hypotheses in front of us: geocentrism and heliocentrism. Although we might lean toward one, as a scientist, we must be neutral when we start examining the two options. But remaining neu-tral is difficult. This is a good time to discuss personal biases that may impact our scientific judgment. If someone makes a big statement, it is quite natural for everyone else to question: "Who is this person?" "Why should we believe him at all?" To answer these questions, we need to know Copernicus better. Copernicus was a church and government official and a physician to some dignitaries. According to today's standards, he was only an amateur astronomer because he did not teach astronomy nor did he have a full-time job observing the positions of heavenly bodies, though he made observations of a few planets' positions using his home-made devices. These facts should not affect one's scientific judgment, but people tend to be affected by them especially when the claim is against conventional beliefs. If the theory of an amateur goes against the theory of a Nobel Prize winner or a famous scientist from a prestigious university, would you believe the amateur? If you held a professional astronomer position at the time, would you believe or simply discard Copernicus's new theory? Why should you trust Copernicus and his theory while leading astronomers that you know or have worked with have all told you something different?[2]

We will next learn what led Copernicus to his idea. The heliocentric theory originated in ancient Greece; Copernicus did not invent it. Instead, he cited this old idea to show that he was not totally off the mark. Nevertheless, this idea did not draw much serious attention before him. According to *Calendars through the Ages*, Copernicus started circulating an outline of his theory in 1514. Historical records indicated that, in 1533, his theory was well received by church officials, including Pope Clement VII and Catholic cardinals. Although the situation might have changed, as speculated by historians, after Pope Paul III succeeded Pope Clement VII in 1534, Copernicus was still invited in 1536 by Cardinal Nikolaus von Schönberg to communicate his discovery. Copernicus published his theory in his book *De Revolutionibus Orbium Coelestium* (*On the Revolutions of the Celestial Spheres*), *before* his death in 1543. These details were significant because some historians and philosophers of science misrepresented the book as found in his will, making an impression that Copernicus

---

[2] A similar example is that of Gregor Mendel (Moore, 2001), who was also not a scientist but a priest and taught physics in his abbey. He conducted experiments on plant hybridization in the garden of his monastery. Less lucky than Copernicus, his results (1865–1866) were ignored by the scientific community at the time because he was not a scientist. Long after his passing in 1884, his results were rediscovered in 1900 by three independent duplications that confirmed his results, which became the foundation for modern genetics. Mendel is now known as the father of modern genetics.

expected a "revolution" due to the publication of his book. Incidentally, Copernicus used the word "revolution" in the title of his book to mean "moving around" and not to mean reversing social orders. His theory became controversial only some 70 years later when Kepler's and Galileo's works became available, during a period of religious conservatism influenced by the Protestant Reformation and Catholic Counter-Reformation. Therefore, the controversy more likely resulted from political and/or religious changes than a scientific revolution. The controversy surrounding the theory might not be a purely scientific problem or even a philosophy of science problem.

Now the question is as follows: If you were a scientist living during that time, what would you do? You might conduct research and give public lectures to either support or be against Copernicus's theory. Copernicus's theory was proposed after the global voyages of Christopher Columbus to the west (1493–1503) and Ferdinand Magellan-Juan Elcano's circumnavigation (1519–1522). During these two voyages, astronomical observations had been used as a fundamental navigation method because the geomagnetic field direction had been unknown at the time in large areas of oceans. Therefore, the positions and motions of the Sun, Earth, Moon, and some planets were tested during these voyages. It would have been noted if the geocentric model had been entirely wrong. At the time, you might assume the geocentric theory could make adequate predictions, as it was tested and verified extensively during these and other voyages. Therefore, at the time, you may conclude that there was not much observational evidence for Copernicus to propose a new idea against geocentrism.

Why did Copernicus think that the commonly accepted idea was wrong? What was his evidence? Were the predictions of geocentrism wrong? Or were heliocentrism's predictions more accurate? Let us examine his reason. In the preface of his book, the basis for his new theory was *not* about geocentric predictions being inaccurate — there was no evidence for it as I explained above — nor heliocentric predictions being more accurate. Instead, he questioned the "modifications" used to improve the predictions of geocentrism. This may *not* sound like a strong reason for challenging the fundamentals of a long-established theory because it is a common practice in science to modify a model in order to improve its theoretical predictions with observations. For example, when someone proposes a model, others may find it interesting and use it to make additional predictions. If the model's predictions are not very accurate, others may find ways or mechanisms to improve its predictions. After a few modifications, the predictions may become very good. Then, people will conclude that this *original model* (with these improving modifications) must be fundamentally correct, i.e., it describes the reality. This was the situation that Copernicus faced. He questioned the modifications introduced over time for various reasons, although all improved the geocentric model's predictions. Copernicus argued that the reasoning for some of the modifications conflicted with one another, i.e., in Copernicus's view, these modifications were not consistent. Describing these different assumptions, he famously stated, "A monster rather than a man would be put together from them" because the hands, feet, arms, and legs were put together in random ways. One may be puzzled by this argument as the basis of a scientific proposal because there are plenty of useful models that have been developed this way, especially in scientific applications or engineering. This is simply how science moves forward; some may argue against Copernicus. Shouldn't Copernicus have based his theory on scientific evidence, for example, reduced errors or improved predictions?

Nevertheless, he argued that one should base a theory on a single set of assumptions to predict the motions of all five planets known at his time. We will learn later in this book that this is a correct requirement for science. However, he thought that these basic assumptions should be as follows: (1) The observation of the planets should use a homocentric method instead of using eccentrics or epicycles, and (2) the planets obey the principle of uniform motion. Today, we know that both assumptions are incorrect because all planets move in ellipses and the orbital speed changes in different parts of each orbit as described by Kepler's laws. He based his theory on seven specific assumptions (although it is a single set of assumptions, seven is not a small number). He conducted detailed analyses of locations for a list of stars, a few planets, the Sun, the Moon, and some astronomical phenomena. I note that Newton's laws of motion did not exist yet.

Given so much information, it is your turn to make your judgment on which theory, geocentrism or heliocentrism, could result in you being on the right side of the unknown history of the future.

As a scientist, you may first question the assumption or hypothesis that the Earth is moving and search for evidence supporting or against this motion. You may be able to do so "deductively". If the Earth circles the Sun once per year, you find that the Earth moves at about 30 km/s around the Sun using the current astronomical unit. However, the distance between the Sun and Earth appears to have been underestimated at the time by a factor of 62. Therefore, the Earth would need to travel at 480 m/s if the heliocentric theory is correct. This is an awfully high speed! Is there any evidence for such a high speed of the Earth's movement? Maybe or maybe not.

Let us test it. Although Galileo's famous free-fall experiment on the Leaning Tower of Pisa was done more than 70 years later, you might think of a similar experiment. But let us use the free-fall experiment as an example to examine whether the Earth is moving at a high speed, an unfortunate omission in most teachings about heliocentrism or Galileo's work. If the Tower of Pisa is 56 meters high, and you drop the ball from the seventh floor out of eight, the height is about 50 meters. After you release the ball, the ball would fall through the air for 3 seconds before hitting the ground. Galileo would have recorded this 3-second time lapse as a part of his experimentation. During the 3 seconds, if the Earth travels in the horizontal direction at 480 km/s, the tower will move 1.45 km or about 0.9 miles horizontally circling the Sun. This means the ball should hit the ground nearly a mile away from the point just below the tower. This is measurable. As we know, the ball would hit the ground vertically right below the dropping point as predicted by the geocentric model. If you were an unbiased scientist at the time and followed the scientific method strictly, you should conclude that there is no evidence for the Earth to circle the Sun. Therefore, Copernicus's theory is wrong.

Furthermore, the (modified) geocentric model predicted the location of the planets more accurately than the heliocentric model did during this time because Copernicus assumed the planetary orbits to be each a circle at a constant speed, none of which is true as we know now. If you were a scientist at the time, which theory would you believe? Are you going to be able to make the correct judgment? You have the new hypothesis of heliocentrism and your evidence is derived according to the H-D method. The results are undebatable: (1) The ball-dropping experiment does not show evidence for the Earth's motion and (2) the heliocentric model does not predict the position of planets more accurately. If you were following the advice from the scientific method provided by conventional theories of philosophy of science, it is no surprise that you would be against Copernicus's theory because his hypothesis is not supported by the available evidence[3] and the prediction is not better. I recall that the evidence for his argument was the inconsistencies among the modifications of the geocentric model, which is not a strong scientific argument because making improvements to a model is a practice widely used in science. Therefore, if you have followed the "scientific method", you would be on the wrong side of history. This is a nightmare that has haunted all theories of philosophy of science so far (Feyerabend, 1975).

Fortunately, Galileo did not follow the scientific method. There is a famous passage in the *Two Chief World Systems* in which Galileo says, through his mouthpiece Salviati, that we should make reason conquer the senses:

> *You wonder that there are so few followers of the Pythagorean opinion* [that the earth moves] *while I am astonished that there have been any up to this day who have embraced and followed it. Nor can I ever sufficiently admire the outstanding acumen of those who have taken hold of this opinion* [Copernicanism] *and accepted it as true: they have, through sheer force of intellect done such violence to their own senses as to prefer what reason told them over that which sensible experience plainly showed them to be the contrary.* (Galileo, 1967)

Galileo recognized that evidence from experience does not support a moving Earth. Rather, he developed the Principle of Galilean Relativity and a theory of inertia, which eventually became Newton's first law of

---

[3] Copernicus's theory faced another challenge called "stellar parallax", which occurs when an observer moves. Under this situation, one sees the motion of a nearer object relative to a farther object. If the Earth moves around the Sun, one should see a change in the position of a close star relative to a more distant star. This parallax effect was not observed at the time of Copernicus and Galileo because it was too small to measure with the instrument available. The effect was first observed by Friedrich Bessel in 1838. Nevertheless, the absence of parallax had been accounted as a serious negative result to Copernicus's theory.

motion. The logic presented above, when arguing for the Earth not moving, assumes that after the ball is released, it does not move horizontally while in the air. According to Galileo's new idea, when one releases the ball from the Tower of Pisa, the ball carries the horizontal velocity of the Earth's motion. Therefore, it moves horizontally at the same speed as the tower. Therefore, even if the Earth is moving, the ball should still hit the ground just below the dropping point.

In this example, you find that a single hard fact that the ball hits the ground right beneath the dropping point can be used as evidence for either the Earth moving or not moving. As I will discuss throughout the book, even if the evidence is used to make the final judgment of an idea, the *evidence* itself may depend on interpretation and may not be the hard fact itself.

The final turn or conclusion of this story may be surprising to most philosophers of science. Galileo stated, "I might very rationally put it in dispute, whether there be any such centre in nature, or no." That is, there is no "center" at all — neither the geocentric nor heliocentric! In other words, geocentrism versus heliocentrism is not a real science problem but more of a religious or political problem because both theories are *correct* according to the future Newton's theory. According to Newton's theory, it is just a matter of choosing a specific "frame of reference", i.e., a center that is assumed not moving. Choose *any* center you wish! This conclusion has been overlooked by philosophy of science and society at large when most people consider heliocentrism correct and geocentrism wrong. To a scientist, in order to answer the question of whether there is evidence for geocentrism or heliocentrism, one needs a different way of thinking. As will be discussed in Chapter 9, scientists "take" the heliocentric theory as a "better" model because it is *simpler*. Scientists and philosophers of science again have different perspectives.

1.5 Science Revolution: Invention of the Quantum Theory* (Elective)

The discovery of the quantum theory is a famous and interesting example to learn about how science revolutions can take place. Because the process of this event has been described incorrectly as an example for many theories of philosophy of science, I will provide a slightly more detailed description for you to appreciate the cause of a scientific revolution. Some of the concepts of philosophy of science discussed may become clearer when studying Chapter 6.

There was a so-called blackbody radiation problem: When measuring the sunlight's intensity as a function of the radiation's wavelength from space, one obtains a smooth curve, without spikes or notches in the visible band, similar to the blue line in Figure 1.1. This type of spectrum is called blackbody radiation. The term blackbody needs to be defined more carefully, but for our purpose, it is sufficient to neglect the spikes and notches if it is observed on the ground. There is the Stephan-Boltzmann law (1879–1884) that states that blackbody radiation depends only on the temperature of a blackbody. Therefore, measurements of the radiation spectrum of a body can be used to derive its temperature. Wilhelm Wien proposed a mathematical form describing the blackbody radiation spectrum (1895) based on Kirchhoff's law (1859), which is called Wien's

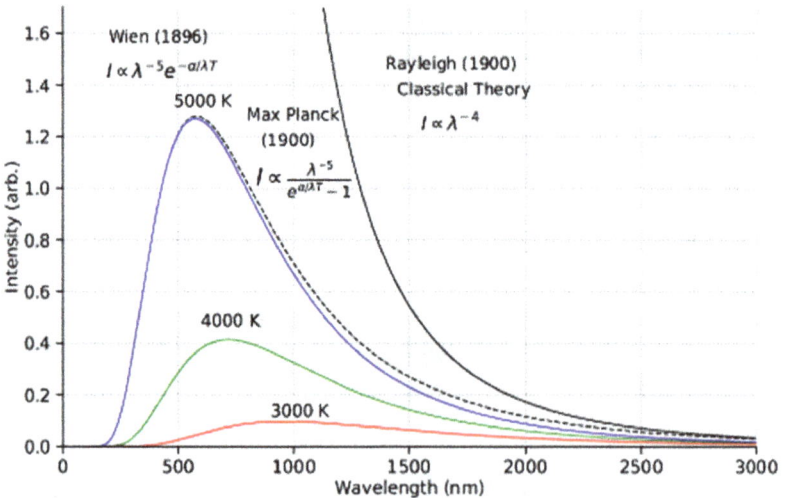

Figure 1.1. Blackbody radiation. Red, green, and blue lines are the radiation based on Wien's approximation for 3000K, 4000K, and 5000K, respectively. The black solid line is given by the Rayleigh–Jeans law. The dashed black line is the formula proposed by Max Planck that bridges Wien's approximation and Rayleigh–Jeans law. The mathematic expression for each model is given.

approximation today. Figure 1.1 shows the radiation curve for a few temperatures as labeled. The formula describes blackbody radiation well, especially at the wavelength of peak power as a function of the blackbody's temperature, which is known as Wien's displacement. Wien's displacement has been observationally confirmed and has been taken as a natural law. For example, the frequency with the highest power of sunlight is in color green, as discussed in Section 10.6, with a blackbody temperature of about 5800° based on which we derive the surface temperature of the Sun. But the formula does not do a good job in longer wavelengths. In long wavelengths, the radiation can be better described by the Rayleigh–Jeans law (1900–1905) which was based on the theory Rayleigh proposed (1869) explaining the blue sky. Planck found a different formula in 1900 that bridges the Rayleigh–Jeans law for longer wavelengths and Wein's approximation for shorter wavelengths. I have provided the mathematical expressions for Wien's approximation, Planck's formula, and Rayleigh–Jeans law in Figure 1.1. One may be able to convince oneself mathematically that Planck's formula approaches the Rayleigh–Jeans law in the long wavelength and Wien's approximation in the short wavelength.

Let me summarize what happened here: We have a single observation (blackbody radiation); there are two theories that can be used to explain it. Each theory had solid observational evidence for itself. However, neither theory can describe the blackbody radiation well enough across the whole wavelength range — each theory is good for one side of the observation. Someone (Planck) came in and introduced a mathematical expression to describe the radiation in the whole wavelength range without rigorous explanation for its reason — either theoretical or observational — but the expression works well to bridge the two theories. On the surface, it is more a mathematical maneuver than an earthquake in science! There was no appealing reason, i.e., no rationalization, which conventional theories of philosophy of science discuss to believe that the new mathematical expression described reality or experience. It does not seem to be significant either.

However, a scientist does not stop after inventing a new mathematical function that seems interesting. Rather, they try to understand what this expression can tell them. For example, how can the constants in Planck's law be determined? This may seem a special effort to non-scientists; it is very common to a scientist who has invented an idea. The investigation led to the invention of Planck's constant[4] in the same year. After Planck's constant was invented, it became clear that the exponential in the denominator of Planck's law is the ratio of the energy (density) carried by the radiation to the thermal energy of the blackbody. The energy of the radiation is then related to the radiation frequency via Planck's constant.

However, this presents a new problem. When the frequency goes to zero, the radiation goes to infinity. This is a serious problem that does not occur in either Wien's approximation or the Rayleigh–Jeans law. This problem can be avoided if the frequency is not allowed to be zero, which implies that there is a required minimum amount of wave energy. In this case, the radiation may be understood as comprising many small energy packages. Then, these energy packages of light are called "photons", which is what Einstein proposed in 1905. The smallest packages of energy are now referred to as "quanta" (or a "quantum" when individuated).

From these new ideas, a whole paradigm, quantum mechanics, was formed in the 1920s. It can explain blackbody radiation, as discussed above; however, blackbody radiation describes only a "smooth" radiation spectrum. There is radiation that appears as notches and spikey lines, which are called absorption lines and line emissions, respectively. Line emissions can be easily understood by the quantum theory as photon emissions of the same amount of energy from the same type of atoms. The absorption lines of a solar spectrum that is observed more easily on the ground are due to molecular absorption when sunlight passes through the atmosphere. Both emission lines and absorption lines are material dependent and have been known since the early 19th century. The thin lines in a spectrum should have been taken as a hint for the discreteness of photon energies. In retrospect, line emissions and line absorptions should have prompted scientists who studied the spectrum of sunlight to propose the quantum theory.

---

[4] The Planck constant is $h = 6.6 \times 10^{-34}$ J Hz^{-1}. The energy of a photon is $E = h\nu$ where ν is the frequency of the photon. The frequency of the visible light is centered at about 550 mm. The energy of a photon in this frequency range is $\sim 10^{-19}$ Joule, an extremely small amount of energy!

The impact of the quantum theory is far-reaching. The invention of semiconductors and semiconductor technologies may be the best-known example of the applications of quantum theory. Can you make a connection between computers or smartphones and the blackbody radiation spectrum?

Questions for Thinking

1. Everyone has a collection of stories and events about themselves. Think of a past event concerning judgment that had to be carefully thought through, i.e., not in a rush, nor under pressure. Without knowing the consequences and current view of the event, what event do you consider to be the most memorable good decision made in your life so far? Now, assess your knowledge of its consequences and your current view: Would you have made a different judgment? What have you learned from it?
2. Answer the previous question, but for the decision that you regret most.
3. Are there any logical flaws in the paradox of the ravens?
4. How would you judge Copernicus's theory during his time?
5. What have you learned from the example of geocentrism vs heliocentrism about personal bias when making judgments? How do you make judgments on whether to believe a website?

Chapter 2

What is Philosophy?

2.1 Why Should a Good Scientist Learn Philosophy?

Albert Einstein once said (1944),

> "So many people today — and even professional scientists — seem to me like somebody who has seen thousands of trees but has never seen a forest... This independence [from prejudices] created by philosophical insight is — in my opinion — the mark of distinction between a mere artisan or specialist and a real seeker after truth."

A scientist should be a truth seeker and not a mere specialist! It is widely accepted in the science community that philosophy, i.e., the ability to see the forest, is essential, especially to the leaders of each science discipline.

By its nature, scientific research involves scattered pieces of incomplete information. A scientist has to find useful and relevant pieces of information, from a mountain of potentially useful information to a problem, and make sense of them with reasoning. Since the problem can be complicated with intertwined observational, theoretical, mathematical, and technical issues, no one would know what the correct answer should be or even if there is one at all before a study. How can a scientist know what to do (*how to get it right*)? And how can a scientist know whether a result

25

obtained is correct or not (*how to know I am not wrong*)? Every scientist will encounter these two questions to a certain extent. People who do not plan to face them would have to work through their whole career under a supervisor who is patient and available at all times. In this case, they would not be scientists but rather artisans or specialists as Einstein referred to.

A phenomenon that has been commonly observed is that great scientists, such as Einstein and Stephen Hawking, all became philosophical. Many scientists become more interested in philosophy when they are involved in establishing a fundamentally new theory or proposing a major change to the existing theory, i.e., when the science discipline is in crisis (e.g., Kuhn, 2012). As we will learn in this book, this is because the scientists encounter the question of how to get it right and how to know they are not wrong. They have to spend much time convincing themselves that the theories existing before their time are problematic and the new theories they propose contain no substantial flaws.

A good scientist needs to have deep technical knowledge and understanding of a broad range of science subjects and is able to critically assess what the most important problems are and what problems are potentially solvable at a given time. A leading scientist needs to be able to determine what the key missing information is and whether it is obtainable, given the available intellectual power plus financial and technical support. They must have a vision that can guide the community forward. When making such assessments and defining the future, knowledge and sound understanding about critical technical points are crucial while detailed technical or mathematical knowledge becomes less important, if the leading scientist appreciates the roles each technicality plays and the intellectual potential of individual scientists in solving these technical problems. It is nearly impossible for a non-expert, such as a philosopher, to conduct such a task. A scientist who is narrowly focused and only enjoys the neatness and beauty of technical details may have difficulty making sound judgments on diverse alternative approaches. Therefore, the task requires that the leading scientists of a field know not only technicalities but also philosophy. In this book, I present perspectives of philosophy of science that are relevant to scientists and scientific research while providing guidelines that address the concerns raised by Einstein.

2.2 The Origin of Philosophy

To study philosophy of science, one has to first learn what philosophy is and what it was developed from. Philosophy may mean different things to different people. To many people, it is when debates are filled with xx-ism, yy-ism, and zz-ism, or -nian. Today, philosophy is defined as "the critical examination of the grounds for fundamental beliefs and an analysis of the basic concepts employed in the expression of such beliefs" (Encyclopedia Britannica), or the "study of general and fundamental problems concerning matters such as existence, knowledge, values, reason, mind, and language" (Wikipedia). The first definition is general but may be too general with several loaded terms for many people to appreciate it so that ordinary people may not be able to understand what it is about. The second one, with a laundry list, is more specific though it is still difficult to pinpoint what philosophy is about. A philosopher, Kasser (2006), defined it more interestingly: "philosophy is the art of asking questions that come naturally to children, using methods that come naturally to lawyers."[1] According to his idea, philosophy is to ask a big question and then divide it into many small questions to answer. This captures a certain aspect of philosophy well, but what is philosophy really and what was it developed from?

Let me decipher it for you. In many places on the internet, one can find that the term *philosophy* is derived from a Greek word meaning "love of wisdom". So what? Who else does not love wisdom? Further reading finds Greek words "sophia" and "sophist", mean wisdom and, literally, wise man, respectively. After more research (scientists, as well as philosophers, do not simply read what was written but also question what was not written), you would be puzzled by the fact that these wise men were those whom ancient philosophers actually argued with and debated against. Therefore, a simple interpretation of philosophy as "love of wisdom" does not make sense because philosophers disagreed with the sophists who

---

[1]This definition may have some similarities with that of Bertrand Russell (1918) — "The point of philosophy is to start with something so simple as not to seem worth stating, and to end with something so paradoxical that no one will believe it" — although it is more confident with its conclusion than Russell's.

already possessed wisdom. And, from reading, it seems that at the time when philosophy was in development, sophists were the more respected scholars with more prestige, and philosophers were underdogs. This might be why the philosophers of ancient times, as documented by Plato, cheered so much when Socrates won a debate against a sophist. With these facts and arguments, the situation became understandable to a scientist like me: These, sophists and philosophers, were two types of schools with different ideas about what and how to teach. Each was trying to attract the brightest students. For sophists, it is clear that they taught "wisdom" so their students became wiser, gained more respect, and could find better or more important jobs. Of course, they would also charge higher tuition. Philosophers, on the other hand, according to their name, seemed to be in an inferior position because they seemed not to possess wisdom but only love wisdom, or "wannabe" wise men in today's language. Who wants to enroll in a wannabe good school and not a good school today if the tuition is not an issue?

It is commonly agreed that the practice of Western philosophy started with Thales of Miletus (624–546 BC) and that the word "philosophy" was first used by Pythagoras of Samos (570–495 BC). Socrates (470–399 BC), who was written about by Plato (428–348 BC), was a renowned philosopher in ancient Greece who had several debates with famous sophists of his time. These debates displayed the full character of the philosophers. During these debates, he first played ignorant on the subjects, letting the sophists explain their wisdom. He then started questioning every point stated by the sophists, found inconsistencies or flaws, and eventually took down their theories. You may now understand the real meaning of "love-wisdom" — being modest and not claiming to know anything. This method is referred to as the Socratic Method, which is also known as the "dialectical method of inquiry". As will be explained and discussed in Chapters 7 and 9, this type of reasoning is fundamental to conducting science.

Additionally, it seems common that these ancient philosophers paid more attention to the natural world, or what we would call natural science today, whereas sophists emphasized the art of words or rhetoric. I would interpret "philosophy" as teaching and learning that included natural sciences while sophists were teaching and leaning more toward language and skills of using language, which are more important for obtaining a position in government. After twenty-four centuries of dominance by the philosophers, this history has been rewritten and philosophy has lost the original meaning and has been interpreted literally as love-wisdom. It is arguably true that sophism is still being taught today in law schools, although no one wants to admit to being associated with sophism.

Socrates was lucky to have a student named Plato who in turn had a student named Aristotle. These three giant thinkers, together, set the foundation for Western philosophy. Socrates, Plato, and Aristotle discussed a large range of fascinating subjects (e.g., Bartlett, 2008). For example, they argued that the ideal government would be governed by a philosopher king, not by democracy. The Allegory of the Cave is a famous metaphor for knowledge in society, according to which our knowledge (in ancient times) could be very limited and distorted, or maybe even irrelevant to reality. They discussed in-depth justice, virtue (What is virtue? Can it be taught?), ethics, politics (a famous statement of "human beings are political animals"), laws, happiness (Are you happy when you have money and fame?), friendship (How to know who your true friend is?), etc. Aristotle (384–322 BC) was a scientist, by today's definition, in physics, astronomy, geology, biology, and psychology. He is the most influential philosopher in history. Interested readers should be able to find many interesting books to read on these philosophers and their philosophy; the time would be very well spent.

Socrates famously questioned the power of Gods and their ability to produce thunder and lightning, which were thought by ancient people to be a result of Gods' anger. Careful observations indicate that thunder and lightning are most often correlated with thick clouds that block the view from/to the heavens. With this information, Socrates questioned the ability of Gods to produce storms, in a way that we now call correlation study in science. Sadly, this was a crime at the time because of impiety. For this and another reasons, he was sentenced to death. I think observations like these, in combination with a connection to the natural world (or science by today's definition), set philosophy apart from sophism.

It is commonly agreed that the central difference between philosophy and sophism is the method of inquiry, i.e., the Socratic or dialectical method which involves criticism and self-criticism. Socrates is famous for saying, although a little bit too radical for me, "an unexamined life is not worth living". This is not about the subjects of study, not about a particular view on a subject, and not about who is right. In philosophical debates, students can question and challenge a teacher's ideas. When one can question everything, nothing is above reason. A consequence is that one can question God or the existence of God, as described above. However, it is somewhat strange to a scientist that, on record, Socrates won every debate except the last one when he was found guilty and sentenced to death by vote. The history appears to have been selectively recorded, a scientist would suspect.

2.3 Philosophical Thoughts in Other Cultures

It is interesting to note that during the same time period of the development of Western philosophy, other civilizations developed Eastern philosophical thoughts, such as in India and China. For example, the Upanishads (Vedic Sanskrit text of Hindu philosophy) were written in 600–300 BC, Gautama Buddha lived in 558–491 BC, and Confucius (who is known as Kong Zi in Chinese, where the first word of the name is the family name and the second word, "Zi", is a respected title for a scholar) lived in 551–479 BC. They were contemporaneous with Pythagoras who lived in 570–495 BC, but lived prior to Socrates. Of course, the information propagation speed at that time was very slow; these philosophical ideals should have been developed independently. Like Western philosophy, the philosophical developments in China and India were also based on open debates.

I once had an interesting interaction with an enlightened Buddhist monk. I learned that to get enlightenment, the brightest monks have to go through a five-year individual study, which is comparable to the average time taken to earn a Ph.D. in America. We discussed the predictability of the future and incarnation in front of a large group of Buddhist students in a temple he was in charge of. You might be curious about how it went when a scientist met an enlightened monk. In general, we were able to communicate without problem; however, we reached different conclusions. For example, he did not believe that we were able to predict whether it would rain or snow tomorrow. This is understandable given that the weather in the high-altitude mountain area is rough and difficult to predict. A question the Buddhist monk asked me more than once is "Aren't you worried to become an ant in your next life?" Ants are among one of the worst realms of rebirth. When he heard my negative answer to the question repeatedly and firmly, the monk seemed very puzzled and in disbelief.

The history of Chinese philosophical developments is very interesting. During the period of Confucius, the central government of China was either extremely weak or nonexistent and thus allowed all kinds of ideas to flourish. This era is referred to as the period of hundred schools of thought. Confucius's teachings were more concentrated on the "problem of conduct" and the "problem of governance", but less on the "problem of knowledge", the three problems of philosophy to be discussed in the next section. He thought knowledge of nature, i.e., science, involved *skills* or *tricks* that could be useful for artisans but not worth studying by scholars. There were also sophistic teachings (famously championed by Hui Zi), a school of military theories (famously edified by Sun Zi, the author of *The Art of War*, a popular reading nowadays in business schools), a school of law (famously conveyed by Han Fei Zi), a school of natural science (famously taught by Mo Zi), and schools of Tao, an alternative philosophy to Confucius's (famously started by Lao Zi[2] and Zhuang Zi). People would go to their open lectures to learn from and to challenge the speakers, who could be hailed or booed at the scene. Confucius's philosophy was the most influential because it set a foundation of ethics and rules for how people should behave in families and society, as well as how

---

[2] Lao Zi, a contemporary of Socrates, is widely considered a philosopher by Western philosophers. He developed the highly abstract concept of "Dao", often translated as Tao, which means the underlying principles and interconnections of how the natural world and human society operate. This concept is similar to "truth" in philosophy, as discussed in Section 2.5. Similarly, he developed the concept of "De", a set of characteristics that people emulate. This is similar to "virtue" discussed in philosophy. Western philosophers often credit Socrates–Plato–Aristotle for inventing the concepts of "virtue", "truth", and "justice" in Greek so that philosophy could be developed more quickly in Greek. It is arguably true that Lao Zi also discussed all three concepts together with the concept of "Zheng" (not leaning on one side) for "justice". Arguably, Lao Zi even recognized deductive reasoning.

governments should rule. Qin Shi Huang (r. 221–210 BC), a prominent figure in Chinese history similar to Alexander the Great in European history, built the Qin Dynasty. He not only conquered other kingdoms but also unified the country with the same written language and standardized the measures of weight and length. However, he prosecuted many followers of Confucius, and burnt most books written, ending the flourishing period of the hundred schools of thought because freedom of thought made it difficult for the government to control society. The Qin Dynasty was built upon the basic idea of the school of law, or legalism.

In the Han Dynasty following the Qin Dynasty, the emperors realized that Confucius's theory could actually be used to help the government because it teaches people to be moderate, obey orders, and minimize confrontations. Due to this, they banned all other schools of thought except Confucius's doctrine. Since Confucius looked down upon natural science as it was for artisans, science was not promoted by this policy. This situation was exacerbated by the national exam system established by the Han and adapted by later dynasties. The exams had three levels, where the final level would be conducted directly by the emperor himself. It is obvious that the emperor would not test the finalists on their ability and knowledge of mathematics or physics. To recommend the best candidates to the emperor, the second, or state-level, exam, focused on governmental policies and strategies, in addition to Confucius's doctrine. Therefore, for almost two thousand years, young people would spend most of their time studying Confucius's doctrine and national policies and strategies. In my personal view, this exam system and the topics of the exams diverted the most talented young people away from science development and eventually slowed the progression of science in China until the end of the last dynasty in the early 20th century. As a result, although the ancient Chinese had many eminent scientific inventions, the early scientific revolution did not take place in China.

Other civilizations that have had a long history, such as Egyptian, Babylonian, and Persian, had their philosophical developments earlier but at slower speeds than that of the Western and Eastern civilizations. Curiously, in these civilizations, there were few prominent philosophers with systematic philosophical ideas that covered a broad range of issues in the early period of their history. Based on this correlation, an argument can be made that the development of philosophy, especially systematic philosophical theories, can positively affect the development of civilization, even though the development may also be affected by many other factors. Of course, the (lack of or slow) development of written language within these cultures may also play a role in this. However, one could argue that if the rulers were philosophical, they should have felt the need for inventing a written language.[3]

The philosophy discussed in this book is mostly based on Western philosophy, as it is more systematic and aligned with our modern scientific reasoning.

2.4 What Does Philosophy Study?

Philosophy mainly concerns three problems: *knowledge*, *conduct*, and *governance* (Robinson, 2004). You may find this characterization radically different from the conventional teaching by most philosophers, such as those discussed at the beginning of Section 2.2.

You may appreciate these three problems to occur roughly in step with the development of civilization. The problem of knowledge concerns what knowledge may be and how knowledge can be acquired. Ancient human beings accumulated a lot of knowledge: Certain fruits are poisonous; tigers, lions, and snakes are dangerous; diseases can kill you; etc. People do not need to understand why something happens in many cases, although often the cause was attributed to misbehavior. But knowledge is the basis for any rational decision to survive. It is usually deterministic. If one is against knowledge, they may be more likely to die. Scientists are more concerned about the problem of knowledge than problems of conduct and governance. It is a question of truth or falsity.

As civilizations developed, more people started living close to each other. Conduct became a problem as one's conduct would affect others.

---

[3] One culture invented the use of strings and knots of different colors to book various items. They made copies of these colored strings and knots on stones, indicating they recognized the need to archive the records as they changed with time. But unfortunately, this culture still failed to invent a written language or numbering system for this relatively simple but obviously needed task.

Distinct from the problem of knowledge, the conduct problem is about *making choices*. When choices are present, the issue is not about truth or falsity, but about which decision is good or better versus bad or worse. An extension of this would be toward questions such as what is beautiful and what is ugly or what is more economic and efficient. The choices individuals make are not deterministic because people view things differently.

When populations become larger and more concentrated, how to govern becomes a bigger concern since people may have different behaviors and interests. Rules would have to be set and judgments would have to be made in order to avoid society's self-destruction from an escalation of conflicts. The fairness of a judgment may be questioned because people may have different opinions about what is preferable and best for individuals and society. There are also some common expenses related to governance, including those for defense, infrastructure, large hydraulic/irrigation projects, and public services. There are several forms of governance, and each form has some advantages and disadvantages. In principle, there are three widely recognized types of governance: by a single person (autocracy), by groups of people (aristocracy), and by a lot of people or the whole of society (democracy).

Today, each of these three problems has evolved into several relatively specialized fields of inquiry. Subjects such as natural sciences focus on the first problem concerning knowledge. Many developing science disciplines, such as economics and psychology, primarily involve the problem of conduct. Most problems assessed by engineering disciplines are not about finding truth or falsity, but rather question the efficiency, economic effectiveness, safety, and reliability of various systems and their components. The distinction between science and engineering is raised here because some existing theories of philosophy of science were based on engineering examples, leading to fundamental confusion in these theories. Engineers may use scientific thinking to address their problems although the problem concerns mostly how to decide among multiple options, i.e., the problem of conduct. In these cases, there are only better choices but no best or correct choices because people evaluate options based on a set of standards that are more or less subjective and weighted with personal opinions and interests. Since the goal of science, in contrast, is to invent new knowledge and to correct errors in existing knowledge, scientists have to make judgments between alternatives related to the problem examining the validity of each piece of knowledge. This, however, is not a problem of conduct because knowledge itself is deterministic eventually.

The theory of science presented in this book focuses on the problem of knowledge. Because many problems of conduct are now becoming fields of science, the theory described in this book is applicable to the disciplines or subdisciplines concerning aspects of the problem of knowledge versus aspects that primarily focus on problems of conduct. Whether the theory can be applied to problems of conduct requires further investigation.

The most striking phenomenon in philosophy, to most people, may be the labeling of an idea as an "-ism". This labeling is supposed to be a way of clarifying and distinguishing an idea in debate. Such a label, however, appears to have some magic power that makes the idea *exclusive* and *universal*; one has to either submit to it or be against it as a whole. With these labels, the philosophers involved may radicalize or uncompromisingly generalize the idea universally, to everywhere at every time and under every condition, until they make flawed statements. Every theory has its limits and applicable conditions; simple labeling often results in the rejection of useful ideas or distorts the original idea. I will discuss some examples in this book. The practice may have deviated from the original ideal of philosophy. From the view of a scientist, this labeling effort can mislead people to debate about something other than the original theories. The labeling may be a main cause of turning away many scientists from philosophy. Often, a scientific problem is so complicated that one may not even know what the full problem is or if there is a solution at all. Under these circumstances, whether an idea is universally true may not be an issue.

To a scientist, philosophy often means to follow Einstein's ideal of looking at the big picture, the forest, not only the details, the trees. According to Robinson (2004), philosophy is about "how to get it right". I strongly agree with his characterization of philosophy — simple and clear. One may argue that this question has been, in principle, addressed over the past two millennia and philosophy now is trying to answer specific questions. This is not true. The question or problem that should have been addressed but has not been discussed in most theories of philosophy is "how to get right for everyone" because there may be no unique answer if people have different perspectives. When people have different ideas about how to get it right, how should judgments be made? Should these judgments be made by a national or a committee vote or decided by the president? Even with a vote or presidential decision, there is still the more general question about how we know the choice is right. One way to get it right is via debates or the dialectic method, as Socrates did. However, to my knowledge, there is no widely accepted theory in philosophy to address this problem, especially how to keep the dialectic method productive and under control.

Nevertheless, I find that philosophy has generally evolved away from the original ideal of philosophy. Instead of figuring out how to get it right, it has often focused more on "what is right" or "who is right", which I think is a more specific question than that for philosophers. When one finds an answer to a question, how does one know the answer is correct? Or, could one be fooled by oneself? This is a more important question for philosophers to answer. Without addressing this more general question of "how do I know I am not wrong", discussing specific or practical philosophical problems may not be useful to a scientist unless one can be sure that the specific problem is correctly identified and the solution can be trusted. Searching for answers concerning these two questions, i.e., how to get it right and how to know an answer is not wrong, will be the central theme of this book.

Related to science, significant debates in philosophy have been about metaphysics, which was introduced by Aristotle who wrote the book *Physics*, followed by another entitled *Metaphysics*, meaning after/beyond physics. In the latter book, he tried to understand what is behind the natural world to answer the following questions: "Does nature have to follow laws?" "Must every effect or phenomenon have a cause?" The discussion involves three topics: ontology, cosmology, and epistemology. Ontology (Greek word for the study of *being*) in general, from a point of view of a scientist, allows people to choose one from two contradicting axioms, "materialism" or "idealism". An axiom is a statement that is taken to be true universally. Based on a philosophical axiom, one can derive a self-consistent theory (containing no contradiction). This type of situation is comparable to some problems in science. For example, one can ask whether two parallel straight lines will or will not intersect, which are two distinct axioms. From the former, you could derive Euclidian (i.e., planar) geometry and the latter non-Euclidean (i.e., on a non-Euclidean surface, e.g., spherical or hyperbolic paraboloid surface) geometry. Philosophy of science conventionally does not discuss matters of ontology and metaphysics, a wise decision. Cosmology studies the chronology of the universe which now has become a discipline of physics. Philosophy of science is a subfield of epistemology which will be discussed in Section 2.5.

In philosophy, many debated problems or questions are more ambiguous and subjective to a scientist, e.g., can science explain everything or answer all questions? Because there are infinite things and questions while there will be countless scientific investigations in the future, to debate such a question is not productive. At a minimum, "all questions" should be replaced by "all scientific questions". Then, there is an issue of what accounts for a scientific question. Therefore, the original question is a wrong question to ask from the perspective of a scientist.

2.5 What Is Epistemology?

In philosophy, the theories and discussions that concern the problem of knowledge are called epistemology. Epistemology, from the Greek term meaning the "study of knowledge/understanding", is a major discipline of philosophy that is also referred to as "theories of knowledge" and concerns the nature, origin, and scope of knowledge focusing on epistemological justification and rationality of belief.

A popular question in epistemology is how people acquire knowledge. This seems an easy question to answer nowadays: They go to the internet to search for something. It is true that one can find nearly everything on the internet. However, how do we know if what we find on the internet is correct and can be counted as knowledge, and not just personal opinions? Are there more reliable ways to acquire knowledge? To answer these, we first need to know what knowledge is.

Knowledge: There are many types of knowledge, which are difficult to classify. There is a great deal of debate on the subject that we will not get into. Conventionally, there are three types of knowledge: *knowledge-that, knowledge-how,* and *knowledge-understanding.* Knowledge-that refers to knowledge about facts, such as facts about geography, history, and exotic animals. These are individual pieces of knowledge and do not necessarily invoke the understanding of the relationship among the pieces. One can simply recite them and not need to answer the question of why. Note that observation of something for the first time, i.e., discovery, is often knowledge-that, e.g., a new species of animal or a new phenomenon. Unless many pieces of knowledge-that are connected in an understandable manner, it is difficult to use isolated knowledge-that for prediction or to explain why things are as they are. For example, the fact that the sun rises every morning so far, alone, cannot be used to predict whether the Sun will rise tomorrow.

Knowledge-how refers to the knowledge of skills such as swimming, playing a ball game, skiing, or riding a bicycle. One can possess these skills without understanding how it works, although the understanding of how it works may help one learn knowledge-how.

Knowledge-understanding is different from the first two types of knowledge and is what we commonly refer to as "The Knowledge". This type of knowledge is not merely a record of multiple facts but also contains the proper arrangements and connection among pieces of facts, or knowledge-that, e.g., their cause-and-result relationship. It is also not merely knowledge of a simple skill. Knowledge-understanding can be easily shared across disciplines and may be used as a basis for inferring other things and predicting the future. Science mostly concerns this type of knowledge. Unless specified, the word "knowledge" refers to knowledge-understanding in this book. However, knowledge and science are different; we must wait until the second half of this book to be able to better discuss their relationship in detail. For now, suffice it to state that science is a process to invent new knowledge and filter out false knowledge. It is knowledge in development.

Some philosophers argue that interpersonal skills are an additional category of knowledge. I do not think this category is necessary because interpersonal skills are based on the knowledge-understanding of common human nature and/or the personality of a specific person of interest in a particular situation. With this understanding, one may still have difficulty devising and implementing a correct strategy to an interpersonal issue because it may involve knowledge-how.

We note that our focus is the "problem of knowledge" and not the "problem of conduct". Because there is no option in the problem of knowledge, we do not deal with "prescriptive" or "normative" problems where scientists have to make moral or value judgments for what it "should be".

Knowledge is what a human being wants to gain and is the basis for most wisdom. It can be gained in many ways and by many means. In addition to direct experience, an efficient way to acquire knowledge is to study in school or to read books. Books are a primary source of knowledge; teachers and professors help you understand the knowledge in books, force you to do homework, and eventually test you on whether you have gained the intricate knowledge of the subject. Nowadays, you find that the internet may be a more convenient way to acquire knowledge because you can easily find narratives about almost everything. Learning from the internet has a specific problem: Not everything on the internet is knowledge. Information on the internet could be misrepresentations, biased opinions, or even rumors and lies. Of course, there are also electronic versions of books and peer-reviewed journal articles on the internet, which should be separated from general websites. The above, however, concerns only learning existing knowledge by individuals.

Inventing New Knowledge: Acquiring knowledge in science concerns inventing new knowledge that does not exist yet, i.e., it cannot be found anywhere. As the theory of knowledge, epistemology is a broader field of inquiry than philosophy of science because science is only one of the ways people can invent new knowledge. Philosophy of science concerns mostly how knowledge gets into science books in the first place, e.g., how new knowledge is invented.

To a scientist, a key problem is that acquired or invented new knowledge can be wrong or a mistake. How do we know it is not? This question is particularly difficult to answer when the information is incomplete. We know, for sure, that much knowledge from ancient times is mostly wrong.

How do we know that people in the future will not say the same thing about the knowledge of today? We will discuss these questions in great detail throughout this book because they are important in science and to someone who wants to discover or invent new knowledge. To appreciate these questions, prior to answering them, we need to first define some concepts discussed in epistemology.

Truth is one of the central concepts in epistemology that philosophers have debated on over thousands of years without consensus. I will define it from the point of view of scientists: It is something that is known to exist and cannot be false. Facts and reality are truth for any scientific investigations, but the interpretation of the facts and reality are not, but often confused with, truth. Truth is objective and independent of human beings' recognition, belief, and knowledge. There may be underlying connections among facts. These underlying connections, known or unknown, are also truth and motivate human beings to understand and describe. Science is a process to describe the underlying connections. Note that "truth" in epistemology is different from the "truth" that people are referring to in everyday life, such as, "the truth is ..." or "the truth of the crime is ...". Most often they refer to personal interpretations or beliefs.

Natural laws are humans' descriptions of the underlying connections of the natural world. They are knowledge-that, when describing an isolated phenomenon, or knowledge-understanding, when describing underlying connections among different phenomena. For example, the falling of an object toward the ground is a fact or truth, while the concept of gravity and the gravitational acceleration is knowledge-that. The universal law of gravitation makes connections to many other facts and is a human description of the truth — knowledge-understanding. Similarly, the reception of a light signal from a distance is knowledge-that. Its underlying truth is electromagnetic wave propagation. The electromagnetic theory involving several laws is a human description of the truth. In science, natural laws may also be *referred to as the truth* if they correctly and robustly reflect and are consistent with general facts and reality in a variety of settings, but they are *humans' descriptions and beliefs*.

Truth is external to and independent of an individual's opinion, point of view, experience, education level, wealth, or reputation. Truth may be relatively independent of time, e.g., in the time scale of human civilization.

Truth can often mean something on two different levels: relative truth and absolute truth. The absolute truth, or The Truth, is an idealized concept: We know it exists, but its specifics are unknown; it is what we are pursuing, for example, the underlying connection of many natural phenomena. It is the goal or motivation of scientific research. Relative truth is something that we know to be true based on our direct or indirect experience. It may involve our (but not individual) knowledge and understanding, e.g., the shape of an electron.

I once listened to a philosophy course in which the professor said that an objective of philosophy is to look for truth and that one needs to use reason to derive truth. When one cannot derive truth from reasoning, due to a lack of knowledge, the professor continued, one should ask experts for the truth. These are very confusing statements about truth. It seems that the truth he referred to is relative truth. However, in general, one cannot *derive truth* purely with reason because truth is objective and independent of human beings. Often one may be able to *learn knowledge* from experts. Knowledge is not truth because knowledge can be wrong, but truth cannot be. Nevertheless, to ask experts for knowledge (not truth, according to our definition) may be good advice to the general public but is not useful to a scientist. All scientists derive their results (note, not truth by our definition) with reasoning, but their results may contradict each other. If numerous experts hold different views on a subject, whom should the general public ask or listen to?

Belief is, in contrast to truth, something that one person or a group of people *thinks* is true and not false. It involves personal opinions and is subjective. The concept of belief in philosophy is different from what is commonly used in everyday life and is not about religion. "Common beliefs" are subjective but can be widely agreed upon. As discussed in Chapter 1 and to be discussed in Chapter 3, knowledge is a kind of common belief and not a truth.

In science, it is essential to distinguish physical truths from personal beliefs, as the latter often carry bias, which are something scientists want to avoid at all costs. On the other hand, "truth-bearing beliefs" are important for a scientist to continue pursuing an idea or approach. For example, before people knew that the Earth is a sphere, a flat Earth was considered an unbiased true belief. After people discovered that the Earth is a sphere, the idea of a flat Earth is false. However, the idea of a flat Earth still reflects the fact that the Earth is so large that its curvature is not readily experienced in our everyday life. When one is inventing a new piece of knowledge, especially an important one, other scientists will try to falsify it before accepting it. Easily giving up a truth-bearing belief is also an enemy of scientists, especially when everyone else is arguing against it. Sometimes, belief is in a form of intuition or gut feeling, instead of a hypothesis that philosophers of science often theorize about.

For a scientist, beliefs can change, sometimes very quickly, especially when strong evidence disproves a belief unambiguously. An important trait for a good scientist is being able to start working under a different (or new) belief without forgetting the one that they have recently given up. For example, as discussed in Section 1.4, the result of the free-fall experiment would be against the heliocentric theory. However, Galileo did not give up the heliocentric idea. Instead, he questioned the interpretation of the result and invented the concept of inertia. As research progresses, understanding also changes. Often, rejected ideas can be modified to have a second life.

Justification and Falsification are to show that a belief is consistent and inconsistent, respectively, with the truth. In science, we often use the word "verification" or "proof" instead of justification. The most important thing to note in science is that justification may or may not be relevant to the truth. In the example of the paradox of the ravens discussed in Section 1.3, using a white swan or red coat to prove that all ravens are black is irrelevant to the problem even if it is based on deductive logic.

In science, falsification is at least as important as justification. Often, justification/falsification may depend on the level of understanding an individual has; one's justification/falsification may include flaws and can be falsified. They may also depend on personal experience, methods, and environmental conditions. For now, it is good to remember that one person's justification can be another person's falsification. You may start wondering how science could handle such a big mess of complicated possibilities while producing reliable knowledge. We will discuss this problem in many chapters of this book, so that we may not be fooled by ourselves as well as others.

2.6 What Is Philosophy of Science?

Philosophy of science may be considered a subfield of epistemology, a relatively new discipline in philosophy. It was formed during the relatively peaceful period between the First and Second World Wars by a group of scholars whom I would call "philo-scientia" philosophers/linguists. Their objective was to bridge the gap between science and philosophy. The composition of the original group and their expertise shaped the field in both good and bad ways. They lived during the exciting period of revolution in physics when Einstein's relativity and quantum mechanics were just born. Most wisdom-loving people, scientists or non-scientists alike, were very confused at that time. Newton's theory was once thought to provide a complete understanding of the natural world, but it seemed to encounter fundamental problems when a body's motion and/or size approached an extreme, e.g., when the size is exceptionally large or small. Philosophers of science have often labeled this deficiency as the "fall" of Newton's theory (e.g., Godfrey-Smith, 2003), a widespread, but flawed, description. Similarly, the wave theory of light has been often cited as an example of false theories in some conventional theories of philosophy of science (e.g., Kasser, 2006). Nevertheless, these philo-scientia philosophers thought that the imagined massive failure of classical physics was due to the lack of a "scientific method". This led these philo-scientia philosophers to try to define and develop a better foolproof version of the scientific method.

Most scientists disagree with the overall assessment, as evidenced by the fact that Newton's theory and Maxwell's electromagnetic wave theory remain among the most important part of university-level science and engineering curricula. These conventional theories of philosophy of science should have explained why high-rise buildings remain standing and airplanes are still successfully shuttling people and material items around the world and how scientists can continue inventing new technologies based on Newton's theory and the wave theory of light.

Unfortunately, some influential fundamental theories of philosophy of science were developed based on these problematic assessments by the philo-scientia philosophers/linguists, not by physicists or scientists. Clearly, philosophy of science has been driven without major input from the science community, as will be discussed in Chapter 4. For example, on Wikipedia, one says that "philosophy of science is a sub-field of philosophy concerned with the foundations, methods, and implications of science. The central questions of [philosophy of science] concern what qualifies as science, the reliability of scientific theories, and the ultimate purpose of science. This discipline overlaps with metaphysics, ontology, and epistemology, for example, when it explores the relationship between science and truth." This definition may describe well the problem of the foundation of the traditional philosophy of science that I will criticize throughout this book: Many of these listed issues do not concern scientists, while most issues that concern scientists are not on the list. It is somewhat surprising that the general philosophical view of science and scientific research discussed by Einstein of seeing both the forest and trees is less discussed in existing theories of philosophy of science.

For example, the debate over geocentric versus heliocentric theory has been a frequently used example for right science. Philosophers reached an agreement of choosing the latter. However, Galileo's challenge to the wisdom of choosing an axiom of a common center of the universe, as discussed at the end of Section 1.4, has not been mentioned in any theory. For debates like this, there is no absolute right or absolute wrong because it is not about science. In other words, to discuss the existence of a "center" is not a scientific approach since this is predominately a question for metaphysics versus physics.

Since the new field was driven by logicians and linguists, philosophy of science has not focused on the most important issues concerning scientists. The conclusion that philosophers of science draw from this situation is that "many scientists today take little interest in philosophy of science, and know little about it" (Okasha, 2016). This has been a very unfortunate conclusion based on professional bias; philosophers of science have not questioned the possible cause of the situation — the theoretical development of philosophy of science could have been on a wrong path. I recall that, for a scientist, one of two basic questions to ask is *how to know I am not wrong*? Philosophers of science should have asked this question to themselves.

Despite their well-justified motivation, decades later in a widely quoted and discussed comment, Richard Feynman, a famous physicist and Nobel Prize Laureate, said, "philosophy of science is as useful to scientists as ornithology is to birds." This was a very strange comment. People disagree on its exact meaning. Some argue that ornithology is useful to birds since it can help protect the birds and others argue that Feynman did not know philosophy well enough. To a scientist, however, this comment is clear enough. When I first read about it, I marveled at Feynman's ability to concisely describe the problem. Firstly, this comment focuses on philosophy of science, rather than philosophy itself. Secondly, philosophy of science does not address questions that concern scientists. The analogy is clear: Birds live their lives without caring about how ornithologists (human beings that study birds) think about what they are. People who have argued ornithologists help protect endangered bird species think that philosophy of science can protect endangered science fields. Otherwise, the knowledge gained in these fields would be lost. Even if some science fields are endangered, philosophers of science should not and would not be able to protect them. Different from an extinct bird species where the bird species cannot become alive again, the documentation of an extinct science field may still be preserved in archives because of the cumulative nature of knowledge. When needed, i.e., if there is real value to scientific research, some scientists would be curious enough to find useful information from historical records, like how we learn about our past from history and archeology. These ideas can become alive again just as Copernicus did with the ancient idea of heliocentrism, which was considered a completely dead idea at the time.

But, if philosophy of science is so useless to scientists and its theories are so flawed, why am I wasting my time writing this book, one may ask? As I discussed in the previous sections, philosophy of science is essential to a good scientist who needs to see the bigger picture, or as Einstein stated, to see both the trees and forest. Its theories, however, need an overhaul by including the perspectives of scientists. A fundamental difference between a scientist and a philosopher of science, especially when discussing an example in philosophy of science, is that the philosopher assumes that the correct answer to a question is already known based on knowledge long after the debate was settled, such as in the case of geo-centrism versus heliocentrism (but they still managed to get it wrong — as we know, there should be no center). On the other hand, a scientist is seeking the most likely correct answer among multiple potential answers to a scientific problem at the time with incomplete or inaccurate informa-tion. In other words, the scientist is living in the time when a debate is ongoing, and most philosophers of science are living in the future after the debate is settled. Scientists each have to make their judgment with incomplete knowledge and information, just like a bird to find worms or build a nest, while not worrying about if ornithologists think whether they are doing the right thing. Philosophy of science is useful if it can guide this decision-making process. Alternatives and other possibilities should always be in the mind of a reader while learning a theory in phi-losophy of science.

Philosophy of science should investigate philosophical questions that arise from reflecting on science. These questions should be shared among multiple branches or disciplines of science and the answers should not be specific to only a subfield of science. Philosophy of science can help achieve the goal of science to invent new knowledge, which leads us closer to the truth with minimal possibilities of making systematic mis-takes. Philosophy of science should be able to explain how science could continuously produce new knowledge to update or remove outdated knowledge. A theory of philosophy of science should provide guiding principles for scientists to make judgments among multiple potential theo-ries with only partial information. We are not concerned about whether the natural world has future goals, purposes, and ends.

In addition to the introduction to philosophy of science, in this sec-tion, I have spent more time raising questions and commenting on the existing theories of philosophy of science rather than providing the answer to the question in the section title. These questions and comments may motivate the readers of this book to investigate the issues in philoso-phy of science. I will discuss some of the answers in Chapter 9 after learn-ing about knowledge, science, and their differences.

Questions for Thinking

1. Are we able to fully understand and describe the natural world?
2. What is science in your opinion?
3. How can you know if the knowledge that you have obtained or find is correct or trustworthy?
4. Socrates once said, "An unexamined life is not worth living". How important is it to be rational? How important is it to have passion?
5. How important is truth and seeking truth? Very often, the truth, especially absolute truth, may not be found anyway, so why bother?

Chapter 3

How Do People Acquire Knowledge?

3.1 What Is Knowledge?

Now, we are ready to define knowledge, which is a collection of justified beliefs that have withstood falsification tests and are held by a large portion of society.

Firstly, knowledge is a belief and not a truth. Knowledge in this book generally refers to knowledge-understanding. It is a human invention to describe and reflect the underlying connections between the natural world and truth. In some sense, knowledge has a common feature with truth, e.g., the independence of individual beliefs. The key difference is that truth is objective but knowledge is subjective, and that truth is reality and knowledge is the description of reality. Knowledge is more like an image of the real world. The quality of the image depends on the quality of the camera that took it and the data processing afterward. A low-quality lens or inaccurate color sensors may distort reality. Although knowledge may be passed down over a few generations without significant changes, it is different from the truth, which does not care about the existence of the human species at all.

In the example I discussed in Section 2.5, the philosophy professor confused knowledge with truth. Like the professor, many people think knowledge is the same as truth. For example, there has been a debate about whether knowledge is "discovered" or "invented". Those who think that knowledge can only be "discovered" take knowledge to be truth itself.

In their view, knowledge, the same as truth, is objective and its existence is independent of humans' recognition so that it can only be discovered or revealed but not invented. However, the flaw with this view is apparent because some pieces of knowledge have been proven to be wrong at different times in history. But truth stays the same and cannot be wrong. We do not know whether the knowledge we currently possess would remain correct two thousand years later. Therefore, knowledge cannot be the truth itself.

In order to clear this confusion, we need to distinguish knowledge-that from knowledge-understanding. Knowledge-that may be similar to truth under most conditions and considered to be discovered because it does not involve human reasoning. Knowledge-understanding, on the other hand, involves more human reasoning that provides connections among multiple pieces of knowledge-that. Some of the reasoning is based on incomplete information and may produce potential distortions from the truth. Therefore, knowledge-understanding is invented to reflect the truth. The possible confusion created by philosophers may be due to the over-simplification of knowledge. It uses examples of knowledge-that to discuss the relationship between knowledge and truth but uses examples of knowledge-understanding to discuss the relationship with science. Therefore, some theories conclude that science equals knowledge which equals truth. For scientists, we are more interested in inventing new and correct knowledge-understanding than in answering the philosophical question concerning the concept of truth — the question which has been discussed extensively in metaphysics over thousands of years without much consensus. We will not further discuss the concept of truth.

Secondly, knowledge is different from science. As will be discussed in Section 8.1, science is a *process* of inventing knowledge; the result of science is knowledge. Science can be understood as knowledge in development, but it is only one of the ways for humans to gain knowledge. Before becoming knowledge, a scientific result is held and understood by a small number of scientists and experts. Nowadays, because of the easy access to information about new scientific results, people have heard more about the newest scientific results and have equated science to knowledge.

However, each of us cannot be an expert on every subject and can only remember so much knowledge. Knowledge must be in a highly condensed form and applicable to many different subjects and situations. Therefore, scientific results have to undergo abstraction and reduction before becoming knowledge. For example, heliocentrism was originally a scientific idea in Copernicus's and Galileo's time; Newton's theory abstracted it to knowledge that provides understanding. Then, through reduction, heliocentrism versus geocentrism became a non-issue and was only an issue of convenient choice. Other than the conceptual sketch of the solar system, details of Copernicus's theory and assumptions are mostly incorrect and did not become knowledge. Similarly, a large fraction of the scientific results is intermediate and is partially correct; the details most likely are actually false. Without a distinction between knowledge and science, examples used in the debates of many problems of philosophy of science are often mixtures of science and knowledge but are used to draw flawed conclusions.

Thirdly, an important feature of knowledge is its social attributes. Even if it could be false or include flaws at a given time, knowledge is shared, commonly accepted, and passed on to future generations. Individual belief is not knowledge. In this sense, although knowledge is a belief, it is independent of the individual. It does not matter whether an individual person believes or not, or whether the person is alive or dead. The independence of the knowledge from individuals may be the cause why some people mistakenly consider knowledge to be objective as discussed above. To become knowledge, individual beliefs, ideas, and opinions have to be justified and accepted on the societal level.

Scientific results are knowledge in the developing stage and may be believed by some but not by others. Most people in society do not even know about them. When the results are tested in the publicizing process, they may be modified or partially rejected. In this regard, unpublished books or notes found in someone's tomb are not knowledge until the contents have been debated, verified, and accepted by society. For example, when Darwin's evolution theory was first proposed, it was not knowledge. It became knowledge after more evidence was found driven by society-level debates. The beliefs held by a few experts, some of which may eventually become knowledge, are knowledge in a science stage, which will be discussed further in Chapter 8.

Knowledge is beliefs held by a large portion of the population within a society, not only by the few experts who have tested it and obtained positive results. The reverse is not true; not every belief that is accepted by many is knowledge, such as "big" lies and rumors. Many advertisers try to make some features of their products sound like knowledge. But advertisements, in general, are not knowledge.

Since we have talked about society, it is worth clarifying that a "society" does not necessarily mean the general public. It can refer to a scientific society or a science discipline, but not a topical group that consists of experts on a special topic of science. There is a question of whether some locally held beliefs, that may have some predictive capability, by a small community are knowledge. For example, some local beliefs are developed based on some rare coincidental events. Because the events are real and occur rarely, falsifications and/or justifications of a belief cannot be conducted adequately. These beliefs may be considered part of the knowledge of the local community.

Fourthly, knowledge has to be testable with experience for its trueness and be able to withstand all substantial falsification tests. For example, theories based on supernatural and spiritual beliefs, such as religions, cannot be tested or withstand falsification; they, hence, are not knowledge. On the other hand, many scientific ideas are not directly testable or there may be some weaknesses in its theory that cannot withstand all falsification tests. These ideas are also not knowledge, yet. I will discuss this issue in Chapter 8 when science is carefully defined.

Knowledge has to be based on and/or describe the experience, even if it is subjective and may or may not correctly describe or reflect the truth at a given time in society. However, because people in society have diverse experiences, to have them agree on a similar set of knowledge is not obviously easy. In ancient times, knowledge was verbally transmitted by "sayings". Today, the standard form of knowledge may be books. Because our society has been and is still producing a great number of books of various qualities for people to read, the current knowledge is included in commonly used authoritative books and textbooks, not every book. The reason for commonly used textbooks being an objective form of knowledge-understanding is that these textbooks have been more carefully examined and justified, by instructors and students at the societal level. In contrast, scientific results are often published in journals and reported in news media. For example, Maxwell's electromagnetic theory in his original paper is science and Maxwell's equations in textbooks are knowledge. Nevertheless, scientists who conduct original research will have to read the original scientific papers in order to understand the thinking, reasoning, and inspiration of how the results were derived.

Can knowledge only be found in textbooks? Can it be found from lectures or in lecture notes? My experience may be somewhat surprising: maybe or maybe not. Once when working on a project, a professor sent me his lecture notes on the subject. The notes had been used for more than 10 years by the famous professor, but I found many obvious fundamental errors in them. Therefore, I think lecture notes may not be a good source of knowledge even from a famous professor in a prestigious university, mostly because the intellectual pool involved in examining the contents is small and not necessarily of high quality on this subject. In a different example, I watched a video lecture from a different, but also prestigious, university and found that the instructor made some obviously wrong statements. To my surprise, in this case, a student raised his hand to challenge the lecturer. Clearly, the lecturer has used the notes many times without a problem, but once in a while, there are exceptional students who can find the errors of the professor! I am sure that the error will be filtered out from the notes next time.

Knowledge can also be found on credible websites. However, when the contents of these websites contradict that in the textbooks, the latter should overrule the former because the textbooks have been scrutinized more extensively by society. A large fraction of the websites is not qualified to be knowledge, especially from websites that have commercial or political motivations.

Finally, knowledge is cumulative. Knowledge best reflects/describes truth and reaches closest to the truth that humans may achieve at a given time but may later be modified or falsified, while new knowledge is accumulated continuously throughout history. When new knowledge conflicts with existing knowledge, the new knowledge has to have a deeper understanding with more predictive power and be backward compatible to explain the successes of existing knowledge and the falsifications it withstood.

It is important to point out that knowledge defined as such has a time delay in the justification process. Scientific news is, in general, not knowledge, per se, because a large portion of society does not yet accept it. It may be easier for a result to become scientific news than to become new knowledge. We have heard too many false scientific claims from the news. Any scientist whose important scientific results become knowledge before they pass away would be very lucky. Therefore, a scientist should not expect the results they have derived to become knowledge, especially within their lifetime. Be patient; you should enjoy what you are doing and your contributions to the knowledge of mankind.

As discussed earlier, knowledge is used as a basis for rational decisions. Decisions often have to be made relatively quickly. Therefore, the best type of knowledge is easy to understand and remember. While this is not a requirement of knowledge, this is important for a scientist when their results become knowledge. It would be better if results are stated using simple narratives and/or related to a user through simple examples or analogies. For example, the knowledge of Darwin's theory is "survival of the fittest". Highly specialized scientific results usually cannot reach this level, and hence "new knowledge" may emerge after a synthesis or reduction process that often involves collaborations from multiple disciplines.

In summary, our knowledge, even if it is the best description commonly accepted by society, is a description of reality at a given time that may or may not be ultimately true in the future. In science, we question whether a description is consistent with reality, and, if yes, to what degree. In theory, most scientists believe that human beings can eventually develop a description that is consistent with natural reality; otherwise, we would not have chosen science as our profession.

3.2 Learning

Learning is one of the subjects that psychology and neuroscience study. In psychology, "learning" is defined as "the process of acquiring understanding, knowledge, and behaviors, skills, values, attitudes, and preferences" (Gross, 2001). Acquiring understanding and knowledge in psychology mostly concerns how to acquire the *existing* knowledge that is unknown to an individual. This is in contrast to inventing new knowledge that is unknown to anyone. We will discuss some results about learning from psychology and neuroscience. However, as will be discussed in Section 4.5, psychology is still a potential science discipline in development; there are many different theories. Although some of the results might be considered scientific, others are clearly flawed. I pick the ideas that make sense to me. Because these results, with my comments, are still knowledge in development, one should read this and the next three sections with caution.

In ancient times, knowledge-that was gained directly from observations and experience. It first requires a careful record of the location and condition of an occurrence, as well as careful observation and memory of the phenomenon plus other sequentially or simultaneously occurring phenomena. Knowledge-understanding, on the other hand, is based on the correlations and associations of observations of different types with assignments and rationalizations. Causal relation is one of the possible relations. When no correlations could be obviously found, ancient people introduced spiritual or supernatural reasons to explain many observations. It is very important to remember that pieces of knowledge-that were able to be passed on to the next generation due to the generational overlap of human beings. For prominent phenomena that do not occur too often, such as comets, earthquakes, and volcano eruptions, the idea might not be able to be tested in a society in every generation. The stories of travelers from distant regions and/or pieces of knowledge-that from ancestors may be distorted as they pass from person to person; the distortions made these pieces of knowledge-that far from the truth. For example, that there is a type of large sea animal called whale is truth and knowledge-that. When they became sea monsters in stories, it is no longer the truth.

Some correlations among phenomena may be very strong and clear. In these cases, the relationship was passed on as old sayings or folklores. If a prediction was repeatedly inaccurate, the old sayings would not have been circulated for long or eventually would become myths. Knowledge-understanding is different from biological information, such as human genes and basic instincts, and is able to be passed from generation to generation even with a generation gap. If human beings had a complete gap between generations, like some insects or annual plants, no knowledge-understanding could be passed on to the next generation. One may argue that information could be written down, but how could a written language be invented and how could the next generation understand what was written? Knowledge can accumulate over history and civilization can develop because of the generation overlaps. The reason why I mention this issue is that some theories and the interpretations of some scientific experiments in philosophy of science have forgotten this basic fact!

Today, we acquire knowledge from books and schools. This learning process distances us from direct experience, but we can learn a large amount of knowledge in a relatively short time. For example, in a science book, many experiments and/or observations may be described. Although we do not directly experience most of these experiments or observations in our lifetime, such as the ball drop from the Leaning Tower of Pisa, we use them as examples to explain many things. Books and teachers help organize this knowledge into forms that are systematic and can be easily understood by various readers and audiences.

There are many factors in this learning process. Listeners or readers may gain differently from the same lectures or readings depending on their knowledge of the subject, personalities, personal experiences, habits, and/or intellectual abilities. On the other hand, a single person may learn about the same topic differently if it is taught by different lecturers. Furthermore, classmates may also have a substantial influence on an individual student's learning experience. For example, an interesting phenomenon that many teachers notice is that excellent students often cluster in a single year, even if everything else is the same year after year. My interpretation is that when a few excellent students are in the same classes, they can positively change the dynamics of the learning process.

When including web-based information in learning, a substantial effort to filter out undesirable or untrustworthy material is required, since the internet contains many personal opinions and information with hidden motivations. I often use a signal-to-noise ratio, which is the ratio of useful and correct information to questionable information and is a concept used widely in radio and optical sciences or engineering, to evaluate the usefulness of a website. Websites with a lower signal-to-noise ratio are not worth the time to read or filter.

From trustworthy websites, one may find that the internet is an efficient way to learn "knowledge-that", but less efficient to learn "knowledge-understanding". "Knowledge-understanding" on most websites is often patchy or piecemeal in nature. One has to fit or assemble these pieces of knowledge-understanding into one's existing knowledge structure. To build a solid systematic knowledge structure, books and schooling are still the most effective way. Homework and exams in school are an integral part of education. It creates pressure that may force or motivate a student to build a solid foundation of knowledge. Learning by solely surfing on the web would not have this essential experience of learning in school.

3.3 How Can We Improve Learning?

We all want to improve our intelligence and ability. But how can intelligence and ability be measured? There are many theories of intelligence among which the most popular one is the so-called *general intelligence factor*, or *g-factor* theory (cf. Carroll, 1997). According to the theory, intelligence can be measured by the *g-factor*, which is a combination of several general factors such as fluid intelligence, crystallized intelligence, general memory and learning, broad visual perception, broad auditory perception, broad retrieval ability, broad cognitive speediness, and processing speed. Clearly, many of these general factors are determined by human genetics and the environment that one experienced; these concepts may help you improve your general memory, learning, and ability. There is still a question of how these factors should be weighed relative to each other.

Psychologists have conducted many experiments that isolate different effects in the learning process. They found that humans have many ways of learning, such as unconscious (implicit) learning, conscious (explicit) learning, long-term learning, and short-term (work) learning. They found that people can more easily remember and learn from pictures, especially striking or unexpected ones, than from words. This is consistent with the saying that a picture is worth a thousand words. Pictures and words appear to be stored in different parts of the brain. However, what about mathematics? Is an equation a picture or text? A physical law often has a mathematical expression and a narrative in words. For example, Newton's second law of motion is $a = F/m$; its description is that the acceleration of an object is proportional to the force and inversely proportional to the mass of the object. It is obvious that the mathematical form is very concise, occupying less memory in the brain, although how to use it may be less intuitive. The narrative is longer but directly tells you how to determine the acceleration. For example, in reflection after the invention of Einstein's relativity, people realized that Newton's law overlooked one important issue — how the mass of the object is determined. Intuitively, the mass of an object may be constant, say, if it is defined as the total number of particles contained in the object. But in fact, the mass could also depend on the object's speed of motion, according to Einstein's relativity. Therefore, there is a logical jump in Newton's law. As a result, people had missed the opportunity of inventing relativity before Einstein because of this logic jump. This problem may be more easily identified from its narrative form. This has been used as an example of the need for linguistic analysis in some theories of philosophy of science.

From my teaching experience, students are different due to their background, experience, and habit of memorizing knowledge. Some may like the math form because it is clear and simple with fewer chances of making mistakes. Others may find the math form means nothing and prefer the narrative. In addition, the same mathematical symbol may mean different things in different natural laws; this can be confusing to some students. Therefore, no one form is absolutely better than the other to every person. A good scientist may remember either form and know how to translate from one to the other when needed. For example, one can remember an intuitively correct narrative and then be able to write down the equation according to the narrative. Some equations can be complicated with multiple terms and factors, but most scientists are able to derive it at the dinner table on a piece of paper napkin.

Experiments made by psychologists showed that memory first takes a piece of information into the temporary memory and then later "encodes" it to the long-term memory. When needed, information is retrieved from the long-term memory. Psychologists, in some popular studies, have paid special attention to the possible errors that may occur during this retrieval process. The errors can be caused by hints or implications in the retrieving process and can produce ambiguity, e.g., in witness testimonies.

Memorizing and Retrieving: There is no existing theory about the process of how we memorize and retrieve knowledge at the *knowledge* level, although there may be results at the *science* level. With the above background information, let us now speculate how to improve learning so that one can become smarter. In science and everyday life, the *speed, accuracy*, and *amount* of retrievable information or knowledge, in addition to the *depth of understanding*, are often used to assess whether one is knowledgeable and intelligent; it is not by the total amount of stored information/knowledge in one's memory or the g-factor. The speed and details of retrievable information or knowledge are most likely determined by the way the knowledge is digested and stored. Therefore, the speed to find correctly the needed information depends critically on how information was stored in the first place.

If the brain is like a very large warehouse that stores all knowledge one has learned, as well as all stories one has experienced over time, to retrieve a specific piece of useful and relevant knowledge is like finding a needle in a haystack. Even worse, one may have learned some specific knowledge, like physics and chemistry, a few times in life. These pieces may be scattered in multiple places in the brain. It would be faster if the warehouse is well sorted, organized, and properly indexed. We notice that we often recall knowledge according to keywords or narratives or a picture of a person. Therefore, our brain may have a natural-born but subconscious capability to index or categorize information and knowledge although it is subconscious according to the logic or reasoning that is natural to the person.

However, we cannot consciously *control* how or where one stores each piece of knowledge and information. Can we help where and how we store and organize the information in our brain? The encoding process may be relevant to this question. In some models, the memory starts with sensory memory, then goes to short-term memory, and then long-term memory. Although these models may explain some processes of memory, they may not be used to explain the process related to scientific activities because these experiments have been mostly based on simple tasks that cannot describe the process of inventing new knowledge. For example, most often, a scientist starts with a question that may or may not be directly triggered by a specific sensory memory but by a direct recall from long-term memory. The scientist would think of various possible solutions involving many theories and mathematics, some of which may be retrieved from the long-term memory and some of which are produced at the moment of the activity. One would reach a conclusion of the solution to the problem and memorize it, e.g., with a few hints or notecards on the key steps, while letting the details fade away over time.

Experiments have shown that the encoding process is not a simple recording or copying process of what has been learned. Instead, when the content of information is large enough, the brain tends to make the information in the form of narratives and encodes them into the long-term memory in a process called memory consolidation. Similarly, people memorize a picture in segments. As the memory goes overtime, less important parts and the color of a picture likely fade away gradually (because black–white images need much less memory). Some of the colors may be replaced with a narrative or label, such as "a red sweater". Furthermore, there is evidence that the encoding from the short-term, or working, memory to the long-term memory takes place during sleep, which explains why humans need sleep.[1] This may be comparable to copying information from a computer's memory to its hard disk. Dreams are a by-product of this process, produced during the information transfer in the brain. Although people may memorize things differently, the facts that the encoding process can consolidate and modify the information are very interesting to our problem.

Reconciliation of Memory: To each individual, learning of knowledge or memorizing is a cumulative process throughout their life. However, there is a new question: How does the brain determine where and how to put a new piece of information if a new piece of information is related to some existing pieces of information in the brain? Does the brain automatically make connections among them? If there are multiple records about the same event that are stored in different parts of the brain, when we retrieve, which record are we getting? If there is a sequential story, which record will a new episode be attached to? It is more likely that the individual has

---

[1] In ancient times, without a shelter, sleeping was one of the most dangerous activities when a person can be eaten by an animal or bitten by a snake. Many animals do not sleep as humans do. Some theories ask why humans need sleep even if they could be endangered during sleep. An interesting question.

to reconcile the multiple records into one, or at least set up a link among the records. If one does not do this consciously when each piece of the story was recorded, the person would be considered a confused person because the reconciliation will have to be done at the time of retrieval even if one is able to find every record. This person is unlikely to become a good scientist. Therefore, it is better if one does the reconciliation consciously at the time of memorizing.

If the subject is not an event but knowledge, what would happen if the knowledge learned later in life conflicts with the one learned earlier that has already been encoded in the brain? This is a situation a scientist may encounter because they may have learned a few times in their life about the subject they specialize in. It is very likely that the version they learned in high school is inconsistent with that they learned in college or graduate school. In the learning process, one has to consciously compare the new knowledge with the records that are already stored in their brain. The conflicts between the new one and the existing ones have to be reconciled.

Given the discussion above, a better learner would connect new information with one's existing knowledge structure with conflicts reconciled, so that the new information becomes knowledge-understanding and is stored with connections to existing records of knowledge. Since there is no formal theory describing this complicated learning process, the reconciliation has to be done consciously before the encoding which is a process out of our control. We may actively invoke the reconciliation by deep and thorough thinking during the learning process. One may conduct a careful review of what has been learned. In the review process, one may consciously invoke connections between new knowledge and existing knowledge in one's brain. For new information that is consistent with existing knowledge, little additional effort is needed to adapt the new information to the existing knowledge structure. One needs to pay more attention to the new information that is abnormal to or inconsistent with one's existing knowledge structure, similar to the way when one memorizes an irregular verb conjugation. If the inconsistency concerns only knowledge-that, a simple overwriting may correct it. However, if the inconsistency concerns knowledge-understanding that was encoded earlier, one may need to think deeply over the existing knowledge structure in order to accommodate the new information. If one is able to reconcile the inconsistency, the new information is integrated into the existing knowledge and stored once, which reduces the chance of errors when retrieved next time.

I further suggest, as I did in college, to do the review and reconciliation on the same day of learning the new knowledge. In this way, the well-organized new knowledge may naturally extend from existing knowledge with necessary links and with conflicts resolved. This would reduce potential confusion during retrieval. This knowledge package will be encoded to the long-term memory that night. If one does not review the knowledge learned before sleep, the new knowledge may not be digested and reconciled with links to the existing knowledge in one's brain. The unprocessed information would likely be encoded to the long-term memory when one sleeps. Without the digestion and reconciliation process, the new knowledge could be copied to different locations in the brain without proper connections to existing relevant knowledge. When the knowledge is recalled, one has to sort through the unprocessed messy information, maybe from scattered locations in the brain; the individual would look slower, disorganized, and less intelligent. This suggested learning routine of same-day review and reconciliation is consistent with most experimental results concerning memory theory conducted by psychologists and may result in faster and more accurate retrieval of well-organized knowledge and deeper understanding of it.

Making Connection of Knowledge: As discussed above, in psychology, learning concerns mostly the process of acquiring knowledge that already exists. For scientists, learning is more than a simple acquisition. It is also a process of filtering out pseudo-knowledge or fake knowledge that is circling around in society. This filtering is particularly important to mature scientists. Evidence and reasoning have to be presented while both are scrutinized. This is a process not only applicable to the subject of one's specialization but also to other subjects, such as in everyday news or a seminar in a different field. I cannot overemphasize the fact that a scientist does not believe anything they hear without first questioning it. In everyday life, a scientist may be annoying or even intolerable to a certain degree. A philosopher of science may get this feeling when reading this book. We do not believe everything 100% in a book, lecture, seminar, news, or a result that is interpretive or statistically based. Nothing can conflict with the knowledge the scientist has already acquired. If the new information is intuitively consistent with the existing knowledge structure but conflicts with some existing pieces of knowledge, the scientist would have to adjust their knowledge system to accommodate the new knowledge. If the new information is outside the scientist's expertise but is intuitively suspicious, the scientist would ask questions for clarification. When asking questions is not an option, one would keep the suspicion and test it with future examples concerning the issue. If the issue is important, one would likely use the previous suspicion to question a future speaker on a related topic.

A good practice to improve learning while listening to a presentation is to *plan* on asking the speaker one or two questions at the end. This motivation may make you listen to the presentation more attentively. While listening, you may find yourself constantly modifying and refining your question(s). It is possible that by the end of the presentation, you are convinced or unable to identify an intelligent question to ask. But, because you have tried to make connections between the presentation and your existing knowledge, you would learn more and be able to recall related information more quickly in the future. Many good scientists actually take this approach.

3.4 How Does One Invent New Knowledge?

In Sections 3.2–3.3, I only discussed the acquisition of knowledge that is known to some people, even if it is unknown to the individual or it cannot be deduced simply from the individual's existing knowledge. This acquisition process is important. It has been the focus of most studies in psychology and is discussed widely in society and among educators. However, this is not what "gain knowledge" is about in science. Science is to discover knowledge-that and invent knowledge-understanding that is fundamentally new to mankind. The discovery of a new species of insect or bacteria may even not always be a simple problem of knowledge-that, because it often involves or requires components of knowledge-understanding. The new knowledge has to have something that cannot be found in any textbook or learned from any professor. Therefore, philosophy of science should concern how new knowledge can come about and how to know it is not false. It is not like simply solving a jigsaw puzzle, as many people describe science. Because a jigsaw puzzle usually has a known and definitively correct answer, no matter how complicated it might be, it is not analogous to science. If one likes the analogy of jigsaw puzzles, science may be like a mixture of multiple jigsaw puzzle games with some missing pieces and without a chart for any game. I am glad to see that some of the newest puzzle games include many features similar to science. It is obvious that what we discussed in the last two sections is mostly not applicable to inventing new knowledge.

How do scientists invent new knowledge? To simply learn a lot of knowledge and then apply it to solve a problem cannot guarantee an invention of new knowledge because the invention of this nature may most likely have already been done. When starting a real science project, one does not know what the answer would be or whether there is an answer. We often do not even know what the problem is. For example, if you lived before Newton's universal law of gravity and were asked to explain why a bird can fly, you may be clueless about what the problem is about. To answer the question of how scientists invent new knowledge is not simple.

Let us first look at how scientists work. If you ask a scientist about what their most important common characteristic is, most likely, the answer would be *curiosity*. Although many graduate students may start a science project with an assignment from their professors — not starting with curiosity — they would quickly be driven by their curiosity and amazement toward the natural world as to why things go the way they go. If you want to be a scientist and you have not been as curious, you may try to observe the world more carefully and ask more questions. The question may not be like "what makes an apple fall down to the ground?", as stated in the legend about Newton. It can be "why does the apple not fly into the sky or fly horizontally?" The questions may sound like each other, but the former already implicitly assumed and accepted the given known knowledge that apples fall down to the ground. In Section 1.2, I asked a few questions to stimulate one's curiosity. For example, how do *you* know the Earth is a sphere? The curiosity comes from observations about the natural world as well as from purely intellectual exercises.

Do not try to answer the question, if you do have an answer. Try to use it and see if it works. Or, you may try to explain your answer to see if it makes sense logically. For example, because of *A*, I have *B*, and because of *B*, I have *C*. Logicians would analyze whether the logic is consistent with their forms or contains a flaw. You would not be able to invent new knowledge simply from that analysis. In the falling-apple example, before Newton's theory was known, your answer might have been that there is love between the apple and the Earth. Therefore, the apple falls in love. More likely, in this simple straightforward procedure, you will find more questions. Why from *A* you get *B,* not *B'* or *B"*? Similarly, why is it from *B* to *C,* not *C'*? You may question why a bird can fly. Isn't there love between the bird and Earth? If not, why does the bird eventually stand and build a nest on a tree branch? One would respond as follows: Because the bird has a pair of wings that flap. But how do airplanes fly if their wings do not flap? Is the wing or the flapping more essential to fly? Furthermore, airplanes are very heavy and should not be able to float in the thin air as a boat does on water. You know that as a fact because when an airplane crashes into the ocean, it sinks to the bottom of the water. You may be able to ask more questions about each of the potential answers. The important thing here is a range of possibilities — you do not stay with the same question/possibility to look for the answer. As will be discussed later in the theory of science (note, it is not the scientific method), the ability to recognize alternatives is essential to a scientist. To answer these cascading questions can eventually drive your curiosity. As I have stated repeatedly, a scientist does not start an investigation with a theory or hypothesis (but with curiosity and questions a lot of them) and does not follow by proving it deductively (but constantly changes an idea in order to understand the problem), as described by the theory of the scientific method. Let us continue with the process that a scientist experiences.

As curiosity is the start of most scientific ideas, one finds that there are too many things and questions one may be curious about. It is impossible to find the answers to all or even a small portion of them. As discussed in Chapter 2, knowledge-understanding is based on the connection or association of several pieces of "knowledge-that". One would find that some of the questions or phenomena are related, for example, the stock market going up and down, the city traffic coming and going, and ocean waves going up and down. They could be related. In other words, some curiosities can grow into beliefs. This is when a scientist starts to think of justifying and/or testing whether the belief is true or not.

The level of sophistication of the tests and justifications depends on the subject or science field. Science demands scientific rigor or *evaluation*, as will be discussed in Chapter 8. Now, you decide to use an *analogy* to prove your idea of stock market fluctuations, city traffic, and ocean wave patterns being related. You get some stock market data from the internet, take counts of traffic over time to make a chart, and take pictures of waves on a beach. By screening many examples of the data, you are able to find one case from each of the three phenomena that look similar to you. You have found the period and amplitude for each. You then present your idea, justification, and results at a science conference stating that the stock market and traffic can be described using the wave theory that describes waves on the beach. You even write a manuscript and submit it to a scientific journal. You must be very excited as you were very creative (by the way, this is a very reasonable feeling and you do not need to be shy about it; as a scientist, you are supposed to be proud of yourself). You have had a great idea and proved it with evidence, just as described by the scientific method — the so-called hypothetico-deductive method — that you have learned. The scientific method works!

But soon, you will find that many scientists have different beliefs! They will falsify your new idea/belief at conferences after your presentation and in the referee's reports of your submitted manuscript with their counterevidence, reasoning, and they will challenge your methodology of analogy. You would be more surprised to find that some of their ideas/ comments actually were much more developed, advanced, sophisticated, and, in some cases, they even thought through your idea before but gave up on it. Here, I recall that the heliocentric theory that Copernicus proposed was already nearly 2000 years old when he wrote his book. Most of the professional astronomers within those 2000 years must have had some evidence or good reasons to reject the heliocentric idea, although some of their reasons might not be scientific. If you were living in the time before Copernicus and you were smart, creative, ambitious, and very knowledgeable about astronomy, wouldn't you have proposed heliocentrism? Why not? You should have noted that Venus and Mercury never appear around midnight, implying that they never move very far from the Sun. One has to remember the fact that in every generation there are highly talented scholars and scientists in the world. This was true for the period from Ptolemy to Copernicus. If there were a recipe that guarantees one would derive correct results, the heliocentric theory would have been invented, reinvented, and tested.

Although you thought you had a great idea and gathered some supporting evidence, you only find that you are challenged and falsified by your fellow scientists. The good news is that in science you have the chance to defend your idea. Depending on the nature of your idea and the field you are in, this falsification and defense/rebuttal process can be as long as (or much longer than) the time it took to develop your idea. For example, it took 30 years for the hypothesis that the extinction of dinosaurs was due to an asteroid impact to be accepted. During the falsification–rebuttal process, scientists may learn a great deal about their own original idea and crystalize it, if it bears truth. One may realize that some other observations or theories can be considered as either supporting or negative evidence. Very often, the original idea may be modified substantially, and sometimes, it simply goes in the opposite direction of the original.

Now, if your idea has successfully withstood the falsifications, it is published. Note that more often it would not be the original idea or hypothesis you had. However, except in rare cases of instant hits, most scientists in your field will not have time to read about your great idea. This is because of the overwhelming number of publications these days, not counting an even greater number of un-refereed open publications. Fewer people think about applying your idea to explain other relevant topics. For example, it took 70 years for Copernicus's great idea to be picked up by Galileo and even longer time to be theorized convincingly by Newton.

This may be considered a digestion time in science. Your idea is still in the domain of science and not "knowledge" yet according to our definition because a large portion of society, even in your professional society, does not know about your invention. It takes a much longer time for a successful theory to eventually become knowledge of society, such as what is discussed in common textbooks. Often, a piece of knowledge is formed with a collection of successfully justified true beliefs and theories through abstraction, condensation, and reduction. For your wave analogy idea, you or someone else would write a book or a chapter of a book many years or decades later that describes the processes. This includes the similarity between waves in three different settings combined with elegant mathematical treatment. The knowledge, while explaining the analogy, among others, is able to explain stock market crashes, traffic jams, and devastating waves during storms or a tsunami.

If this discourages you from choosing science as a career, it should not. You should choose to be a scientist not because you, your parents, and other people that surround you think you are smart or good in mathematics. Rather, it should be because you are curious and try to invent some useful ideas for the benefit of mankind. You will enjoy the satisfaction of using your intelligence to solve problems or to answer questions no one else can, an experience you cannot get from any other profession. The chance of becoming like Newton or Einstein, or to win a Nobel Prize, is much lower than becoming a famous singer, as we know that there are countless famous singers performing on stages every night, in many cities in every country in the world but there are only few Nobel Prizes awarded each year.

From the above discussion, you may note that the hypothetico-deductive method, if used correctly, may describe only a small portion of the process. Newton once said, "I frame no hypothesis". All scientists should follow where nature leads us. If one starts with a hypothesis to prove, it is more often that one is misled by the hypothesis. For example, in a large project I was involved in recently, we conducted complicated experiments in space. The results are very difficult to understand. Some scientists in the project have followed the H-D method and made the same hypothesis but with different technical treatments for justification. When the hypothesis encountered a difficulty, each introduced a specific potential scenario for their idea to get around the difficulty. Since the details of the ideas are in conflict with each other, they debated heatedly. However, other scientists focused on the problem and not the hypothesis; they were able to prove that the hypothesis was impossible or irrelevant to the problem so other potential mechanisms had to be considered. Therefore, starting with a hypothesis in research is potentially harmful because it may let personal bias take over. Remember, the conclusion of scientific research can be just the opposite of the starting hypothesis.

3.5 Is Creativity Important in Inventing New Knowledge?

Creativity, according to the dictionary definition, means making or bringing into existence something new. Since science is to invent new knowledge, it is a creative endeavor. However, social media and the general public have overemphasized creativity or imagination in science. This may present a wrong impression about science. Albert Einstein once famously said, "Genius is 1 percent talent and 99 percent hard work", meaning that creativity is not as important as it is portrayed. Whether one is creative or not depends on how creativity is defined. For example, it may be defined as "the ability to find associations between different fields of knowledge, especially ones that appear radically different at first" (Foster, 2015). This seemingly reasonable definition may be more suitable for art or literature but not for science. When we examine how important creativity is, immediately, we notice that creativity in art and science is evaluated completely differently. The judgment for creativity in art is more *subjective* and highly *dependent on culture and time*. There is no absolute standard. However, creativity in science has to be verified by reality. This standard is nearly time-independent and is objective.

A scientist, when encountering a very difficult problem, makes associations of the problem with all possible or impossible fields of knowledge. Many of the associations are radically different at first (or creative as defined by the general public) and eventually cannot work out; they are dropped. On the other hand, as explained in Sections 1.5 and 6.4, very often the idea is not as radical as outsiders portray. For example, Einstein's great idea of special relativity, the assumption that the speed of light is independent of the speed of the motion of either the source or the observer, is exactly what was observed by the Michelson–Morley experiment, as will be further discussed in Section 6.4. The difficulty was to work it out. Therefore, the creativity hailed by the general public is not what in science.

"Creativity is not only the ability to come up with new ideas, but to narrow down those ideas to focus on one that can be elaborated. Creative people in any field come up with new ways of looking at the world–they are constantly asking, 'what if...'?" (Bickmore, 2010). This definition describes better the creativity in science. As will be discussed in Section 9.4, "what if ..." is scientific reasoning called "dialectical reasoning", a restricted form of Socratic method. In science, a single piece of observation may generate uncountable numbers of explanations or interpretations. New ideas and imagination have never been in short supply. A colleague of mine once claimed that he has more than ten great ideas every day. Another colleague responded by saying that the problem for him is that he does not know which of the ten ideas are correct. To fully test these 10 ideas may take 10 years because, during a test, a single idea can easily generate 10 more new ideas. There is, in principle, no end to this cascading process. The most difficult task for scientists is not to imagine and create a great idea, but to reject 99 out of 100 great ideas, which is what Albert Einstein's 1 percent talent and 99 percent hard work is about. One learns about the problem in the process of rejecting a large number of possibilities. Although some of the ideas could appear obviously wrong or irrelevant, should we reject them? Who would think that a falling apple could be related to the motion of planetary bodies and heliocentrism? Maybe just one layer below the apparent contradiction lies the greatest idea. (Newton did not try to answer the question of why an apple fell on his head; he tried to understand the "love" among things in the universe.) Creativity may actually be the ability to recognize some ideas that appear impossible and/or are rejected by others because of their apparent impossibility. This is not the creativity as portrayed in social media.

On the other hand, developing a unique and independent theory for each individual observation cannot be counted as science because knowledge-understanding needs to connect several pieces of knowledge-that. An objective of science is to reduce the number of possible explanations and the number of theories. This is the filtering process mentioned at the beginning of the chapter. Some people think that science is to doubt everything. This may be true but doubt alone is not enough. Science has to provide more definitive knowledge to guide people's decision-making process and has to improve a human's ability to solve problems and make predictions.

Philosophical Thinking in Science: Let me describe what I have experienced in inventing new knowledge. In the science problem that will be discussed in some detail in Section 6.5, we aimed at creating a large framework for space physics. The problem may involve applications of knowledge from several science disciplines, such as plasma physics, fluid mechanics, space physics, astronomy, optics, atomic physics, and photochemistry. Mathematically, it may become a large number of governing equations, all of which are partial differential equations. To solve such a problem definitely needs a lot of creativity. Furthermore, this problem has been studied by a very large number of space physicists over more than five decades. However, with new observations, more new ideas have been continuously proposed.

When we started working on the problem, there were many ideas that we could think of. There is no need to simply "create" more new ideas. We tested some of the ideas but were not successful. These unsuccessful attempts helped me understand the internal connections of the problem. But the project could be stuck even with so many creative ideas. During the agonizing period, some days I would wake up around 3 or 4 o'clock in the morning to think over the unsolved problems from the night before. Since I was awake but did not want to get up at this early time, I could only use my working memory and rely on the knowledge stored in my long-term memory in such a circumstance (the details of this process are important for philosophy of science to theorize, as discussed below). I would recall the problem I had, the equations that may describe the problem, and every factor or term in the equations that may play an important role in the problem. I would have to carefully examine the physical meaning of each term and the relationship with other competing terms as well as various possibilities. Often, I would have to recall the order of magnitude of each effect from my memory so that I could decide the dominant effects to keep. I would also recall some previous approaches that I or others had tried to solve the problem with a focus on the possible combinations of these ideas. After two or three hours of thinking and reasoning, with my eyes closed and doing nothing else, a *qualitative* conclusion may be reached. Then, I would get up and write it down. It would be used to guide the detailed investigations during a new day. Such early morning thinking usually brings some significant progress. However, the goal may still not be achieved. The same cycle — generation of new ideas, semi-quantitative evaluation, and critical review — would take place a few times until the project reaches its goal. I am sure that I am not alone among scientists in doing this.

In the whole process of inventing new knowledge, the sparks of creative ideas are most exciting but appear less important because eventually most of them would not work, i.e., one cannot be sure which spark produces the final result at the time of sparking. Critical reviews of the ideas are most important. Thinking and reasoning without pencil and paper forces one to be more conceptual and "philosophical". Is this the most creative thing I do in inventing new knowledge? Yes. Isn't it like lightning striking me? No. However, this is more like what Einstein referred to as "seeing the forest". Therefore, in my view, the function of imagination and creativity in science, although important, has been mystified and over-rated by the general public. I should emphasize that the process and conclusion described above apply well to me and may or may be applicable to others.

This most important process in inventing new knowledge — critical review — appears, in my case, to be accomplished in the working memory. There are some interesting experiments and theories in cognitive psychology and neuroscience about working memory. These experiments tested the size of short-term memory capacity needed for people to correctly do simple tasks such as memorizing unrelated numbers (e.g., phone numbers) and letters. The results have shown that the size of our capacity for processing information is about 7 ± 2 bits, the most famously cited result by George A. Miller in 1956. Some later theories distinguish the short-term memory, which refers to the function of storing unrelated information, from working memory, which also includes the function of manipulating information in addition to the storage function. In the latter case, the information should be measured by "chunks" and not bits. Therefore, the tests should be on words instead of unrelated characters. The result shows that the size of working memory is about 4 ± 1 chunks

(Cowan, 2001). According to either theory, when the number of pieces of information is greater than this capacity limit, the brain cannot handle it and would start producing errors. As described earlier, my critical review process was conducted purely with working memory and it definitely involves much more than five chunks (e.g., 13 partial differential equations in my case) of working memory, no matter how one defines a chunk. According to the theories of psychology, I should have been confused by a large amount of information in my working memory and should have made many mistakes. Therefore, the existing models of working memory are not applicable to creative thinking in science and new models need to be developed.

Critical Thinking: Through the discussion above, it is clear that there is no established theory that can describe the process of invention of new knowledge although people have highly valued creativity in the process of inventing new knowledge. Nevertheless, creativity addresses only half of the question in science — the invention (of a new theory) or discovery (of a new phenomenon) — but an invention could be nonsense and the discovery could be illusionary. How does one know an invention is not nonsense? Why not? This is a real question for philosophy of science and may be more difficult and more important to scientists but mostly unknown to the general public. A fact is that many big inventions/discoveries often end up as big disappointments. Just recall how many times you have heard in the news that someone from a prestigious institution invented a new drug that cures many types of cancers, but more people still die from the same types of cancers year after year. News reporters, in general, pay much less attention to the eventual failure of an invention/discovery as it does not have as much news value, except for a few that involve scandals of some sort. This bias is referred to as a selection effect in science. If the news reporters paid the same attention to reporting the failures of these big inventions/discoveries, scientists would have been more careful when they called for press releases or made big claims. Although the general public may quickly forget about these false inventions, the science community does not. Colleagues, friends, referees, or rivals learn from these failures, although the information may not be made available to News/Entertainment media.

It is quite normal for a scientist to get something wrong if it is NOT in publications. If a publication is found wrong, in a short time period, the negative impact can be substantial to the scientist who made a false press release because it damages their credibility. Why is that? Don't we allow people to make mistakes? No! It is because a new invention must have some elements about which most people cannot make judgments by themselves. It would depend on a review/referee process done by a few leading experts on the subject, who evaluate the newness and authenticity as well as trueness of the invention. If one made a major mistake, how do we know that the person is not making another mistake? People would be more suspicious about their next big claim. In many cases, no one would be willing to spend their valuable time evaluating the claim. After all, to make a false claim is, and should be, the last thing a scientist wants to do (although occasionally someone may do it for other purposes).

As we see now, creativity may have been mystified and overrated and many reported creative ideas may have been mistakes. Those who make careless false claims may and should pay a high price so that they put scientific integrity above their potential personal gains. The price should be high, although it is not high enough in my opinion. Therefore, people have been actively searching for a scientific method that can serve as a recipe for doing scientific research and avoid the possibility of making false scientific inventions or discoveries. On the other hand, society also needs a credible idea for science education and political decision-making.

Questions for Thinking

1. In everyday life, people often say "the truth is ...". What does "truth" mean here?
2. In your opinion, is there any difference between truth, knowledge, and science?
3. What is the most efficient way to learn in your experience?
4. How large do you think your work memory is — the maximum number of pieces of information that you can handle without making mistakes? If this number is great than 5, what is your explanation?

Chapter 4

Elementaries of Philosophy of Science

4.1 Empiricism, Rationalism, and Transcendental Idealism

Because philosophy of science does not discuss ontology, we will not deliberate on debates relevant to ontology, such as those between materialism versus idealism. Before discussing theories of philosophy of science, we need to discuss where new knowledge is sourced from, a fundamental question in epistemology.

From the discussions in Chapter 3, it is clear that knowledge can be sourced from direct or indirect experience. Direct experience in our discussion includes observation, experimentation, and other signals a person receives and interacts with in the natural world. Folklores, books, and schooling are all based on knowledge derived from others' experiences, i.e., indirect experiences. Are these direct and indirect experiences the only source of new knowledge that was unknown before? A problem is that, although humans have had experiences with falling objects since our ancestors, it was not until Newton that we learned that it is caused by gravity, the attraction or love between the object and every tiny piece of earth and rock below us. Newton developed the knowledge that quantitatively explains how the attraction of objects follows a precise form that is proportional to the reciprocal of the square of the distance between the object and each piece of the Earth. This does not seem to be a result solely of the

75

experiences of any individual nor the sum of all experience of the whole of mankind over the totality of history before Newton. Something else is operating. It seems that humans can also learn from conceptualization and reasoning. Therefore, experience and conceptualization/reasoning may be two sources of knowledge setting two major camps in epistemology for discussions about the origins of knowledge.

Empiricism: In philosophy of science, the most influential group of people is those who think that the "only" or "primary" source of real knowledge about the natural world is experience. They label themselves as "empiricists". They believe that we know nothing when we are born, i.e., our minds are a "blank" slate, which is also referred to as "Tabula Rasa", and we learn everything by experience. In this theory, there is no pre-programmed function of the brain concerning knowledge or how to gain knowledge.

Empiricism can be traced to many famous philosophers and was commonly cited as being founded by Francis Bacon, John Locke, George Berkeley, David Hume, and John Stuart Mill. However, I should add a cautionary note that in its early years, empiricism was promoted against the religious belief that God is the source of everything including knowledge. Due to this, some of the quotes from these earlier philosophers of science may have been interpreted incorrectly as being against rationalism in philosophy of science which will be discussed in the next subsection. Promoting the importance of observation and experimentation is not equal to being against rationality or reasoning. Therefore, some empiricism in philosophy of science differs from what was promoted by some of the earlier philosophers. For example, during his time, Bacon promoted rationality even though there was no established theory of "rationalism" as today in philosophy of science.

There was a more radical subgroup of empiricists who labeled themselves as sensationalists. According to their ideal, everything we learn is from *direct sensation* and the purpose of our thought is only to track and respond to patterns in these sensations. To sensationalists, without sensory data directly from the eyes, ears, nose, and tactile sensations, nothing exists in one's mind. A flaw in this logic is obvious because even if nothing exists in one's mind without sensing, it does not mean that everything in our mind is from sensing. In modern science, this is particularly problematic because many things may exist but cannot be sensed directly. This is the case, for example, with individual bacteria, viruses, molecules, and electrons. Or, sensory information may not be within the frequency range for our eyes or ears to perceive, such as with infrared, X-ray, and ultraviolet light or supersonic or subsonic sound frequencies. Or, an object is too far and too faint to see with naked eyes, like many celestial bodies in the sky. Scientific instruments can extend the range of our perception, making sensationalism fatally problematic. Even without scientific instrumentation, is sensation reliable? Why can people draw different conclusions from the same experience? For example, one's perception of temperature is experienced relatively. The sense of warmness of a finger to water depends on the thing, e.g., hot things or ice, the finger touches before touching the water.

Nevertheless, empiricism historically played an important role in advancing human knowledge and science. It eliminates religion as a source of knowledge because the supernatural phenomena that many religions are based on cannot be sensed directly. However, they need to explain how Newton came up with his theory of gravity if the source of knowledge is only experience.

Rationalism opposes empiricism. In philosophy, rationalism believes that the criterion of the truth is not sensory but intellectual and deductive. Its extreme form argues that *we come to knowledge* "a priori", or we are born with reason and logic through which we learn. For example, when you talk to a baby, they can respond correctly to some conversation, implying that there are some inherent functions of rationality before experience, rationalism argues. In contrast, empiricism believes that all *ideas come to us* "a posteriori", meaning everything is learned from experience. Does the baby have knowledge of the language or grammatical structure, empiricism asks? If yes, what language, English, French, or German, is this a priori inherent function based on, empiricism further questions? Before developing an understanding of language's grammatical structure, which we learn much later, a conversation would be simply a series of sound bits. If this example is still not sufficient to settle the issue, let us look at again Newton's invention of the universal law of gravitation. This law, although rational, cannot come a priori because it did not exist before

Newton. One can see that the extreme form of rationalism has some difficult questions to answer. Nevertheless, rationalism takes reason as the chief source and final test for knowledge.

For a scientist, before making observational comparisons to test whether a theory is correct or not, we need to first check if the theory contains any logical flaws. However, where does this logic come from? How do we know whether our answer is correct? Can't reasoning itself be based on successful experiences versus being born with it? Or, is the logic what facilitates knowledge learned through experience?

In philosophy, theories are often presented in extreme forms that are mutually exclusive as discussed in Chapter 2. If reasoning was based on a successful experience, it would align with empiricism. According to the extreme form of rationalism, however, reasoning is not obtained from experience but is born with; there are certain existing truths born with us and our job is to *reveal* and grasp them. Examples of preexisting truths include logic and mathematics. This may be satisfactory to some philosophers or mathematicians, but not to a scientist. Are there identical preexisting truths in everyone's brain? If so, are they the same as they were 5000 years ago? What about the brains of animals? If humans all have the same preexisting truths when they are born, why do some people make mistakes when learning something, while others do not? Are there multiple forms of logic and mathematics? How do we know we have the correct form? Is there an independent way to verify them?

Proponents of rationalism include some prominent philosophers and mathematicians, such as René Descartes, Baruch Spinoza, and Gottfried Leibniz. Here, I should mention that Descartes's famous quote, "I think, therefore I am", is sometimes or often misinterpreted. On the surface, the statement appears equivalent to, as can be found on the internet, "I am able to think, therefore I exist. A philosophical proof of existence based on the fact that someone capable of any form of thought necessarily exists" or "my existence depends on my thinking of my existence". In fact, these are widespread misinterpretations. Descartes made the statement when arguing against skepticism, for which nothing is trustworthy. When one proves something with evidence, skepticists can question the evidence itself. This chain of doubt may continue forever so that no conclusion can be drawn. Descartes introduced his logic with an irrefutable point that stops doubt, i.e., one should not refute one's own existence when they are debating, and then, reasoning can be built from there.

A scientist constantly faces the challenge of potential differences between observation and theoretical predictions. We constantly ask ourselves, is the theory or observation wrong? Which should we trust? According to Descartes, there is "nothing more in my judgment than what was presented to my *mind*". This statement may be relatively unambiguous, meaning that he took his mind as the final judgment, i.e., rationalism, although this does not necessarily mean that reasoning and logic are something we are born with.

We all know that we may not have the correct judgment every time. How can we find out when we are wrong? Similarly, if scientists all have their own judgments, are these judgments the same? If yes, only one scientist is needed for a science problem. But in reality, they do not have the same judgment. How could science have made remarkable advancements today if scientists cannot agree on something? If there is something that scientists with different ideas can agree upon, what is it? This may be a more important question to discuss in philosophy of science, not about whether rationality is innate or learned.

Transcendental idealism was proposed by Immanuel Kant in his book *Critique of Pure Reason* (1781); how to interpret his idea is still being debated in philosophy. Given the opposite ideals of empiricism and rationalism, it is always interesting to learn different views on the topic that also concerns us. As discussed above, empiricism believes that everything we know is based on and is eventually tested with experience. Rationalism, in contrast, believes everything we learn eventually has to be tested by logic and reasoning. Kant analyzes the problem of why people can learn or experience the same thing but reach different conclusions and understandings, a problem that scientists face most of the time. He distinguishes two different types of real things: the person of interest as a "subject" and the thing of interest as an "object". The experience a person has with the thing (object) depends on the person (subject). The object remains the same and real, but what a subject experiences or learns from it is different; thus the object is separated from one's conception and knowledge of it. Therefore, according to Kant's ideal, the world is material, i.e., as both a subject and object, but concepts and knowledge of the world are ideal. This distinction helps explain the difference among people in learning and experience, but it does not explain how scientists agree on something in order to make scientific progress. Again, the question is what do they agree on?

Kant also raised issues concerning space, time, and causality. These three concepts do not have the properties of a material. We cannot directly see or touch them, but we all experience and are affected by their existence when dealing with or studying any object or process. Philosophers debate heatedly about whether these are real things or not. For example, Leibniz thought space and time were not things, though Newton thought they were.[1] Kant argued that we learned about these concepts through interactions between a person's experience and their preexisting mental structures, which are used to make sense of the experience. Note that the concept of preexisting mental structures is used to avoid the problems with the preexisting truths of rationalism. For example, we learned about the concept of space when we were babies crawling to reach a toy; we might take many tries before reaching it, through which we learned about the concepts of distance and three-dimensionality.

In our discussion, we consider that concepts of numbers, time, and space are beyond the domain of philosophy of science. These concepts have been made confusing enough as they are a metaphysical problem in philosophy. We consider the only common sense of 3D uniform space and 1D forward time without invoking general relativity in which space and time are intertwined.

Kant famously said "theory (concept) without observation (percept) is empty and observation (percept) without theory (concept) is blinded" — what a beautiful way to describe the relationship between experience and reasoning. He thought that there is no method to follow for discovery itself, though methods can be useful in their justification. We will further discuss the importance of various methods of justification in science in

---

[1]Leibniz independently invented the calculus method of integration at the same time Newton developed the method of calculus that utilizes derivatives. The debate over who first invented calculus had been one of the most famous fights over intellectual properties in the history of science.

Sections 5.1, plus Chapters 7, 8, and 9, although we have cast much doubt about the idea of scientific methods.

Skepticism: As mentioned above when discussing rationalism, there is also skepticism which, in an extreme form, doubts our ability to understand anything about nature, denying rational belief and the possibility of knowledge. In a mild form, it emphasizes critical scrutiny. As a scientist, because most of the knowledge we have already learned sounds reasonably consistent with our experience, without a certain level of skepticism, it would be difficult to identify areas where new knowledge may be needed. New knowledge may likely seem suspicious at the beginning to other people. Too much skepticism, however, could result in missing many potential new ideas since extreme skepticism would doubt the possibility of any knowledge. Another function of skepticism, in addition to helping invent new knowledge, is to filter out less likely solutions to a problem. If no perfect solution is found, we can take the best solution at the time while remembering its weaknesses or shortcomings. More importantly, as I have emphasized many times so far, self-criticism, i.e., skepticism toward oneself, is also essential to a successful scientist. To doubt oneself is much more difficult, especially for someone who has been successful most of the time. A scientist who might have received straight A's in school and has been praised for being smart throughout their life must learn how to overcome this overconfidence. This issue is discussed further in Chapter 11.

4.2 Deductive Logic

We have learned that knowledge-understanding is to connect multiple pieces of knowledge-that and that knowledge may come from various sources. The connection of these pieces can be made with reasoning or logic. I briefly introduce the logics commonly discussed in philosophy of science by philo-scientia logicians of philosophy and linguists. The term "logic" comes from a Greek word meaning "possessed of reason, intellectual, dialectical, argumentative". Logic is the systematic study of valid rules of inference without flaws, i.e., the relations that lead to a conclusion based on a set of other propositions (premises). More broadly, logic is the analysis and appraisal of arguments. There are forms of basic logic relevant to inferences: deductive logic and inductive logic, both of which were first discussed by Aristotle.

In general, there are requirements for the relationships between the premises of an argument and its conclusion. Deductive logic describes situations that if the premise of an argument is true, then the conclusion is true, i.e., a truth-preserving logic. This logic works when a more general truth is known; we can deduce the trueness of specific things that are related to the general thing. An example of this logic is as follows: "If all men are mortal (general thing) and Socrates is a man (specific thing), Socrates is mortal." At first glance, the logic of this statement may sound trivial; of course, Socrates was mortal. But, when thinking about it more carefully, you may find that a key issue is whether a general statement is true for everything within its relevant category. For example, the statement "all ravens are black" is known to not always be true as a general thing and cannot be used to predict the next observed raven being black. Another key issue is whether the specific thing is completely included within the general thing, i.e., the specific thing may only have an overlap with the general thing. For example, one may wonder whether Socrates is a man or not. Some people think of him as a god or saint. How do we know Socrates is not a god or a saint? Or, are his parents not gods? In some cultures, saints or sons of gods are immortal. The good news is that in science we usually do not need to deal with arguments like these. In science, all correct mathematical derivations are forms of valid deductive logic under the conditions specified in each operation and step of its procedure. In other words, mathematicians have taken care of the specifics. Thank mathematicians, job well done! Now, we need to handle the general thing. For example, we know Newton's law applies to many physical problems, such as those related to sending a satellite into space (as a general thing). Then, we can use Newton's law to mathematically calculate the trajectory of a specific satellite (as a specific thing) with high accuracy. This is the power of deductive logic.

Law of Syllogisms: Because deductive reasoning is the most important form of reasoning in science and it is truth-preserving, we now discuss it in more detail. First, let me introduce the Law of Syllogisms, which uses two premises to derive a conclusion in the following form:

a. Major Premise and Minor premise, therefore Conclusion, a total of three propositions.
b. There are three terms in the syllogism. Each proposition includes two of the three terms and each term occurs in two propositions. This results in a total of 64 possible arrangements. Each arrangement is called a mood.
c. A term in a proposition can be either universal (to all) or particular (to some) and can be affirmative (yes) or negative (no). Therefore, each term has 4 possibilities which are called figures.

For example,

All B is A and all C is B, therefore all C is A.

It has a fixed form. The first premise is called the major premise and the second the minor premise, and the last is the conclusion. In this case, both premises and conclusion are "universal" and "affirmative" in their mood. For a specific example,

All men (B) are mortal (A) and Socrates (C) is a man (B), therefore Socrates (C) is mortal (A).
　Valid.

B is called the middle term; it has to be in both premises, but not in the conclusion. The relationship among A, B, and C can be easily shown with a diagram, as displayed in the left panel of Figure 4.1.[2] The most important feature in this diagram is that B is completely within A, and C is completely within B. There is no overlap. Therefore, C is completely within A. This demonstrates an essential feature of deductive logic that a valid argument has to go from a more general premise to derive a more specific premise, i.e., it is a "reductive" process. If the logic goes from a more specific to a more general situation, it cannot be deductive.

---

[2] In logic, there is a standard way to describe the possible complicated relationship among the three terms with three overlapping circles, called the Venn diagram. Since in science, we can divide a complex problem into multiple more straightforward steps; we can deal with a two-term problem in each step. We will not discuss the Venn diagram because it produces too many possibilities.

If we change the order, say to, "All A is B and all C is B, therefore all C is A", as shown in the middle panel of Figure 4.1, it is not valid because not all C is A.

For example,

The Moon (A) is a sphere (B) and the Earth (C) is a sphere (B), therefore the Earth (C) is a Moon (A).

This is invalid, where, in other words, we say "no, that does not follow".

In general, if the form "all B being A and all C being B", as shown in the right panel of Figure 4.1, may not satisfy the required conditions or the premises are not true, the conclusion of all C being A may not be true. For example, all bachelors are unmarried, and all unmarried men are bald, therefore all bachelors are bald. Clearly, that does not follow because not every unmarried man is bald, i.e., the second premise is not always true. The conclusion needs to be drawn carefully.

As described above, there are $64 \times 4 = 256$ distinct possibilities, called types or syllogisms, among which the conclusions of only 24 types are valid. Although the complete law of syllogism can be useful knowledge, for those who are not logicians, the 256 types may be too many to memorize, and one may benefit from a table that lists all these types. As we know, we can find ourselves in need of making a split-second decision in accordance with our knowledge and we may not have time to check a table even if we have one. Fortunately, for scientists, we only need to consider a few of the simplest types. The type that is used most often involves 1 premise — a condition — and its consequent, a total of 2 "terms".

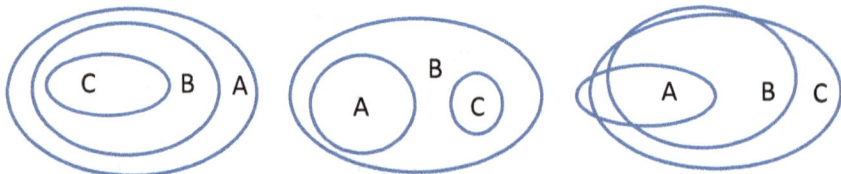

Figure 4.1. Deductive logic involving three terms. There are multiple possible relationships between A, B, and C.

If___, then___ Form is very useful in science and computer programming. It has the form of

"If ___, then ___."

First, we need to specify what the premise and consequent are in the blanks, then we give each item a truth value, which designates whether the specified premise or consequent is true or false. For example,

"If *P (a premise) is true,* then *Q (consequent) is true*".

For whether *P* and *Q* are both true or both false with switching order, we have 2 × 2 = 4 possibilities, or figures. The four possibilities are as follows:

1. *P* is true, *Q* is true (valid)
2. *Q* is true, *P* is true (invalid)
3. *Q* is not true (*Q̲*), *P* is not true (*P̲*) (valid)
4. *P* is not true (*P̲*), *Q* is not true (*Q̲*) (invalid)

Figure 4.2 illustrates the possible situations. The smaller blue oval represents *P* and the larger one *Q*. A premise letter without (with) an underline is true (false). Black (red) labels indicate a valid (invalid) conclusion. Let us take a simple example: If I am in Boston (*P*), then I am in Massachusetts (*Q*).

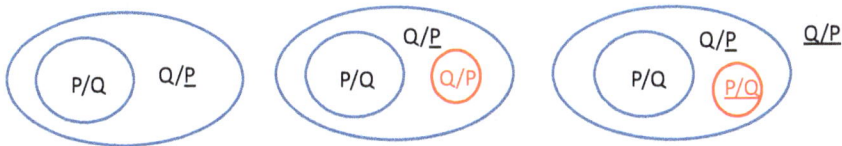

Figure 4.2. Deductive logic form if___ then___. The first label of a pair of labels separated with a slash is the "if condition" and the second label is the "then conclusion". A label with (without) underline is a true (not true) statement. For example, *P/Q* is for if *P* is true, then *Q* is true, and *Q/P* is for if *Q* is true, then *P* is true. Underlined *P̲/Q̲* represents if *P* is not true, then *Q* is not true. Underlined *Q̲/P̲* represents if *Q* is not true, then *P* is not true. Black (red) labels and circles indicate valid (invalid) deduction.

In case #1, *P/Q* is indicated in the left panel of Figure 4.2. This is valid because if I am in Boston, then I am also in Massachusetts as Boston is a city of the state of Massachusetts. We say that *P* entails *Q*.

For #2, if I am in Massachusetts, then I am in Boston, the conclusion may or may not necessarily be correct because it is possible that I am in another town or city of Massachusetts. In the place indicated by a red label *Q/P* with the red circle in the mid-panel, the conclusion is not valid. We say that *Q* does not entail *P*.

For #3, if I am not in Massachusetts, then I am not in Boston. The conclusion is correct and therefore the logic is valid. The black circle of $\overline{Q}/\overline{P}$ in the right panel is outside of the blue ovals.

For #4, if I am not in Boston, then I am not in Massachusetts. The conclusion is not necessarily correct because I am not in Boston, but I could be somewhere else in Massachusetts as shown in the red label with a circle of $\overline{P}/\overline{Q}$ in the right panel,

Soundness: It is a condition, in addition to the "form" for the deductive logic to work. It requires that not only the logic form be correct but also that the truth value or the conditions for premises be correct and that the premises be related to the conclusion. These two requirements are essential to a scientific argument because if the premises do not hold or are not directly related to the conclusion, the conclusion would most likely be false or incorrectly justified. For example, all people can see 3 miles far; a car is 3 miles away; therefore, everyone can see it. This argument is valid according to the law of syllogisms, since "3 miles" is the middle term, and not in the conclusion. However, this argument is not *sound* because the premise that everyone can see 3 miles may not be always true, such as in the case of blind and nearsighted people. Additionally, this premise does not specify the weather conditions affecting visibility; one might not be able to see 3 miles on a foggy day. And, there can also be obstructions between the viewer and the car. A sound scientific conclusion would be that "some people can see 3 miles far; a car is 3 miles away; some people may be able to see the car under certain circumstances". The logic, in this case, sounds very awkward to scientists. It can be rewritten in the form "If ____, then ____": If everyone can see 3 miles away and a car is 3 miles away without any items or conditions obstructing or limiting visibility, then everyone can see it. In this case, the premise, *P*, is "every-one can see 3 miles and a car is 3 miles away without any view obstruc-tion", which are two conditions. The consequent, *Q*, is that "everyone can see it". But, it is possible for *P* and hence *Q* to be untrue, such as in the case of a nearsighted person or foggy day, thus making the argument unsound. This would be the type of logic a scientist often uses and works accordingly. Figuring out the conditions and limitations for the premise to be satisfied is often an essential part of a scientific study.

The premises also need to be related to the conclusion. For example, if a man takes birth control pills and does not get pregnant, then birth control pills prevent the man from pregnancy. The argument is not sound because the man who does not get pregnant has no relevance to whether he takes birth control pills or not. Examples like this are more often used by philosophers and do not really occur in a scientific debate because the flaw can be easily recognized.

4.3 Inductive Logic

Inductive logic works in a way opposite to deductive logic; it makes gen-eralizations, in contrast to reduction. Generalizations are always non-deductive because deductive logic is from more general to more specific. However, it is worth mentioning that using the conclusion from inductive logic to predict the *next* observation is deductive because it is from general to specific. Whether the prediction is successful depends on the quality of the conclusion derived from the inductive generalization but not on the concept of inductive logic itself. In philosophy of science, only deductive logic is accounted as formal logic and inductive logic is suspicious.

Interpolation is one of the most widely used forms of inductive logic in science. It is needed because observations are limited. In an experiment, for example, measurements may be made or tested for a parameter, using values such as 1, 2, 3, ... 100. When we apply the result to explaining and predicting, say for the parameter values between 10 and 20, there are an infinite number of cases that may not have occurred or been tested. Therefore, we would like to generalize the results for the measured condi-tions according to inductive logic. As stated above, whether the prediction is successful depends on the quality and robustness of the generalization because inductive logic itself does not carry prediction capability as it is from specific to general. Some conditions will have to be added when making predictions, as discussed in Chapters 7, 8, and 9.

Extrapolation is also based on inductive logic. It extends an observational result to the parameter range beyond possible observations. In principle, extrapolation often has a limit. Beyond the limit, extrapolation may have more chances to fail (than interpolation) or need corrections. For example, when extrapolating Newton's second law of motion to extremely high speeds, the mass of a moving object is no longer constant. The most useful form of extrapolation is to extend in time, i.e., to predict the future.

It is well known that predicting the future is always risky. Inductive logic is known to potentially produce false conclusions. When using a flawed conclusion to predict, the prediction and actual observation may not be consistent. For example, according to the logic used in the paradox of the ravens, a non-black raven can prove a generalization to be false. However, there is a question of whether the failure is due to induction or due to a prediction based on a flawed conclusion. Prevailing theories of philosophy of science have been developed based on the former, a fundamental mistake, as discussed in Section 4.4.

With these two forms of logic, inductive and deductive, the ancient Greeks were able to derive the conclusion that both the Earth and the Moon are spheres. How did they do it? They first based their theory on observations of lunar phases. Although the Moon sometimes looks like a full disk, its appearance continuously varies, being a fraction of a disk most of the time. When associating the shining part of the Moon with the light from the Sun, the light appears from the direction of the Sun. The logic can be presented according to the two forms of logic: The phases can be produced by a sphere with a moving light source based on inductive logic; one may test the trueness of this conclusion with spheres of different sizes and distances from a light source. Applying this general conclusion to the Moon, a deductive logic, the Moon does not shine its own light but reflects the light of the Sun and the moon is a sphere. Next, if the lunar eclipse is caused by the Earth blocking the sunlight, the Earth's shadow projected on the Moon is a 2D disk in observation. If the lunar eclipse can be observed at a different local time, say in the early evening or early morning, then the Earth is a sphere. This is because if the Earth is a flat disk, the shadow of the Earth in the morning and evening would look more like a plate viewed on its edge. Because the shadow of a sphere is always a circle and the Earth's shadow is a circle, the Earth is a sphere from deductive logic according to the law of syllogisms. The soundness of the premise can be proved based on observation with inductive logic.

There is a mathematical method called "mathematical induction" that can be confused with inductive logic. The objective of the method is to generalize a "mathematical" idea. The method starts by proving the idea to be true with a mathematical expression for the first case which is called the base case. If other cases are in sequence and identical in nature (i.e., can be described by the same mathematical expression), one then assumes the Nth case to be true, to prove the $(N + 1)$th case to be true. If the proof is successful, the "mathematical" idea is proved. This logic appears to be *generalization* (the base case is equivalent to the observation of a black raven) and hence is called "induction". However, the procedure is mathematical and hence is deductive. One may draw a conclusion that induction can be proved deductively. In fact, there are two logical processes involved in this: generalization of the idea and application of the idea to the cases *in sequence which are identical in nature*. This requirement in italic for generalization is rather restrictive, different from *any raven*. Ravens in the world are not necessarily connected or identical. If the $(N + 1)$th case is a non-black raven, it is not identical to the base case. Although the mathematical induction is inductive, the application of the mathematical expression to a specific problem is deductive. In this case, the generalization is the mathematical expression of the idea which is a premise. It cannot be applied to non-identical cases.

4.4 Logical Positivism and Logical Empiricism

In Sections 3.2–4.3, we have discussed general ideas about how we learn knowledge and invent new knowledge. In the rest of this chapter, we will now discuss some conventional theories that have been widely discussed in philosophy of science.

Logical Positivism is the most important and influential theory in philosophy of science because it was the founding ideal of the discipline. The name positivism, although it means something else in philosophy, may be understood as emphasizing the positive aspects of verification, i.e., to prove that something is correct. In contrast, as I will discuss in the next section, falsification can be viewed as negativism since it emphasizes a theory to be wrong. The word "logical" restricts the means for proof because the only acceptable proofs are via logic in this theory. More strictly, only "formal logic", i.e., deductive logic, is acceptable. Later, a moderate version of this theory changed the name to logical empiricism.

It was started by a group of philo-scientia philosophers and linguists after World War I in Vienna and Berlin, who called themselves the "Vienna Circle" and "Berlin Circle", respectively. Again, I use the phrase "philo-scientia" because dominant members were not real professional scientists. Although there were a few *bona fide* scientists, they contributed to some secondary topics. I emphasize this because the founders of the discipline focused mostly on subjects that are only weakly related to science and developed many ideas that appear to be fatally flawed or, at best, irrelevant, as indicated by the large swirl on the front cover of this book. Again, as Feynman described, these philosophers of science studied science and scientists, as ornithologists studied birds. By reading their theories and discussions, they mostly did not know how science was conducted, what scientists needed, or how scientists thought; yet they conceptualized science and scientists in a naïve and biased manner. Their view on science, good or bad, has permeated into today's popular view about science in society.

Philosophically, the philosophy of science movement favored empiricism and opposed the "absolute idealism" advocated by G. W. F. Hegel (1770–1831), who famously stated, "Reason is substance, as well as infinite power, its own infinite material underlying all the natural and spiritual life....", i.e., all of reality is spirit. The time after World War I was also a period of revolution in physics. This was an exciting period in the history of science. Newtonian mechanics had been immeasurably successful and fundamentally changed the world, leading to uncountable technological advancements during the centuries before. Human beings had gained tremendous power and the capability to do most things they could imagine. However, the inventions of Einstein's relativity and quantum mechanics were able to prove that Newton's theory could be problematic under certain situations. From a philosophical point of view, these new theories questioned the fundamental ideas people had formed about space-time and the deterministic nature of science, as will be discussed in

Sections 9.6 and 10.4. Prior to this scientific revolution, Newton's theory seemed to have been proven unambiguously by every non-living thing. How could this carefully and exhaustively investigated and tested theory be wrong? This, unfortunately, was an incorrect assessment because Newton's theory is not wrong but has some limitations that were unknown previously. This erroneous assumption led the field to a wrong path. Founders of philosophy of science argued that there must be some problems in the logic or reasoning in science. Therefore, they suggested that we need to question the "rational" part of epistemology and use empiricism to reexamine how we gain/acquire knowledge.

Because the movement was dominated by logicians and linguists, they carefully examined the logic used in science and tried to identify its flaws. They believed that complicated scientific justifications could be broken down into small logical steps. Theories of language could then make proper definitions of the words so that every step could be fully justified with formal, i.e., deductive, logic. They, following Kant's idea, reasoned that there were two types of sentences: either analytic or synthetic. In philosophy, analytic truths are necessary truths and synthetic truths are contingent truths. For example, the statement "all bachelors are unmarried" is analytic. Mathematics and (deductive) logic are analytic. The trueness of an analytic statement can be determined linguistically. However, although it is true that an analytic statement carries and preserves truth, it is empty truth to a scientist because bachelors are unmarried by definition of the word "bachelor"; no new information is derived. In contrast, the statement "all bachelors are bald" is synthetic. Its trueness cannot be determined linguistically. It has to be proved by other means or it is not true. According to their ideal, science needs only to make sure all synthetic statements are valid.

To help track the logic relationships, they introduced symbolic notations for logical analysis. Symbolic logic and its notations can be used to construct a truth function and describe the relationship of items in a compound statement. Truth tables, which contain values of true or false, under various situations, are now a tool used in many discussions of philosophical and philosophy of science. However, I think that the use of symbolic logic has been a distraction because it overemphasizes the forms of *deductive* logic, rather than the real scientific reasoning of a problem. This has led to many flaws and mistakes in the development of theories in philosophy of science, because in science, other reasoning forms may be needed. I will discuss an example of the fallacy of logical positivism later in this section and another in Section 10.6. I will try to minimize the usage of symbolic logic in this book except later in this section and Chapter 7 where I show that logical symbolic notations are also fallible. The logical analysis used in philosophy of science is ultimately different from mathematical analyses.

In addition to analytic–synthetic distinctions, logical positivism uses the verifiability theory of meaning to understand and regulate science. Verifiability refers to showing something to be true, similar to justifiability. According to it, scientific results need "rational reconstruction" according to theories of language, although nothing is new other than the terminology they introduce. Positivists suggested that science includes two phases, i.e., the context of discovery and the context of justification, again following Kant's idea. As a side note, the notion of context of discovery implies that logical positivism equates knowledge to truth, a flawed usage. If the "discovery" is about a new idea, i.e., inventing an idea, one finds that in science most times there is no way to distinguish the two contexts because the new idea is more often to be found wrong in the justification process, which triggers another new idea or an improved idea. Overall, the analytical–synthetic distinction in logical positivism was based on imagination or rare examples of science.

The Problem of Induction is an acute problem logical positivism encountered in the context of justification. This problem regards generalizations from a single or a limited number of observations to an infinite number of them or from observations to something one cannot observe, such as quantum phenomena. Logical positivists had great difficulty proving that past experiences can be used to predict the future based on inductive logic.

This difficulty was originated by David Hume's argument that "you have no reason to believe that the Sun will rise tomorrow" and by the chicken-farmer problem posed by Bertrand Russell. In Russell's problem, a farmer fed his chickens every morning, so that the chickens, based on inductive logic, predicted that the next morning when the farmer came, so did the food. The prediction was successful until one day when the farmer did not bring the food but rather brought an ax to slaughter the chickens. Similarly, there is also a famous experiment done by Russian scientist Ivan Pavlov who rang the bell each time before feeding a dog. This pattern became so regular that the dog seemed to take the bell ring as a cause of food. Even if Pavlov rang the bell but did not bring food, the dog still reacted as if the food was coming by producing saliva.

These examples illustrate that predictions based on past observations/experiences may be fallible. According to Hume, the prediction is based on "uniformity of nature", although he did not explain what exactly constitutes a uniformity of nature. The commonality of the examples is the regular occurrence of a phenomenon until something else disrupts the regular pattern. Unfortunately, logical positivism interpreted this phenomenon as evidence for the fallacy of induction, a flawed conclusion.

There are also differences among these three examples. Hume's sunrise problem questions the validity of making a prediction based on past experience, i.e., simple generalization of the regularity of a phenomenon based on induction. Because the regularity may be disrupted at any time, the prediction based on past experience is potentially fallible, a correct conclusion. However, we can predict the Sunrise tomorrow, not based on simple past experience, but on knowledge-understanding: The Sun rises each day due to the spin of the Earth which is due to angular momentum conservation. The Earth will continue spinning unless it is hit by a body with sufficiently large momentum from the right direction at the right location to annihilate the Earth's spin motion. This prediction does not involve generic inductive inference or regularity but is based on a natural law. Therefore, yes, we "have reason to believe that the Sun will rise tomorrow". Not only that, but we can also predict the exact time and direction of the sunrise. Although natural laws are gained from experience, they are derived via a restricted form of inductive reasoning as will be discussed in Section 9.3. When drawing his conclusion, Hume oversimplified the problem and did not distinguish between experience and knowledge-understanding. The latter is not simply from experience. His argument and conclusion are fundamentally flawed but have misled almost all investigations of philosophy of science for a hundred years.

In the chicken-farmer case, unless the farmer kills every chicken each time, any survivors of the slaughter (or slaughters if they survived multiple times) would be cautious and react to any sign of danger, thus warning the others, just as the old man in a tribe would do. However, if the farmer slaughters every chicken each time, the knowledge influencing their reactions cannot accumulate among them. Therefore, the presence of a generation gap cannot satisfy the cumulative condition for knowledge as discussed in Section 3.1. Similarly, if Pavlov's dog assessed the situation like a scientist, it may remember being cheated and would not trust the bell ring until seeing food. Nevertheless, Pavlov did not conclude that the dog made the wrong prediction according to inductive logic. Instead, he invented the concept of conditioned reflex in a similar manner to Galileo's invention of the theory of inertia; salivation due to conditioned reflex is a physiological phenomenon and has nothing to do with whether there is a food or not.

Let us examine the cause of the prediction problem in these examples from the view of scientists that have been discussed so far in this book. In all three examples, prediction is made only according to "knowledge-that". We recall that knowledge-that involves a phenomenon that is not connected with other phenomena. When a phenomenon is related to other phenomena, knowledge-understanding can be developed. Logical positivists have correctly questioned the power of knowledge-that as the basis for prediction. However, their conclusion that inductive logic causes false predictions is questionable. In fact, "knowledge-understanding" can be used as a basis for prediction as I explained in "Hume's problem of induction" as well as in the alternative conclusion in the other two examples. Our conclusion would be that prediction can be based on knowledge-understanding. The regularity of a phenomenon is knowledge-that and not knowledge-understanding. It cannot be used as the basis for prediction. If we cannot predict whether the Sun will rise tomorrow, we need to understand *why*. Therefore, the conclusion that inductive logic is not trustworthy in science is a fundamental mistake made by logical positivism. Furthermore, as will be discussed in Chapter 7, any theory of philosophy of science based on the "analogy" of inventing new knowledge to the basic instinct of animals for survival, such as dogs or chickens learning proxies for food, would not be accepted as science because the two processes are fundamentally different.

When examining the failure of a prediction, a scientist would not simply conclude that inductive logic is the cause. Instead, they would conclude that *either* the theory of logical positivism, *or* inductive logic, *or* a specific prediction fails. Using the examples discussed above, which falsify their own flawed inductive inference, is not what a scientist would do. These examples are what many philosophers and I often refer to as "armchair speculations" which are unconstrained arguments framed on epistemology. Therefore, these three examples, in theory, have no bearing on science and hence have no bearing in philosophy of science because they did not distinguish knowledge-understanding from knowledge-that for prediction. With knowledge-understanding, many things in nature may seem to be uniform or regulated, but regularity in observation of a single phenomenon does not necessarily lead to knowledge-understanding. Therefore, the conclusion that inductive logic cannot be used in science is fundamentally flawed. We will discuss how scientists eliminate such potential flaws in Chapter 9.

Scientists do not make predictions based on the "uniformity of nature". They make predictions based on "knowledge-understanding", i.e., not solely based on routinely observed phenomena, which are knowledge-that. Even before a scientific result reaches the level of *knowledge-understanding*, the prediction needs to be made based on *scientific* understanding. For example, with knowledge-understanding, short-term weather forecasts have been improved substantially over the last few decades with better modeling and knowledge-understanding of atmospheric processes, especially the coupling processes between different scales. It is common knowledge among scientists that without knowledge-understanding or scientific understanding, their prediction capability is very limited. For example, no one seems to be able to consistently predict the exact time and magnitude of a stock market crash although rallies may be easier to predict. The reason is most likely associated with human behaviors involved in the stock market and economy. When under stress, the uncertainties in human behaviors can be very large. Humans can react completely differently under extreme stress; we still do not have a good knowledge-understanding of the process. We will discuss the inductive logical processes in much more detail in Chapters 7 and 9.

Therefore, one of the central points of logical positivism — inductive logic is not trustworthy in science — is fundamentally flawed. Unfortunately, the positivists spent more of their effort trying to find ways to fix this imagined problem.

The Paradox of the Ravens, as discussed in Section 1.3, is a famous example invented by logical positivism. Given the problem with inductive logic as discussed above, positivism proposed to replace the inductive logic in inference with deductive logic, such as the H-D method. But it failed — a major failure of positivism. Because it is so important to positivism, let us further examine what causes the problem in this paradox. Figure 4.3 illustrates the situation. The black circle includes all black things, area A. Beyond it are non-black things in area D. All ravens are within the orange circle, areas of black ravens B or non-black ravens C. Now, the questions are whether the orange circle is completely within the black circle, i.e., C = 0, or if it is separated from the black circle (B = 0), or if there is an overlap between the two circles, B ≠ 0 and C ≠ 0, which is shown in Figure 4.3. The hypothesis is to prove C = 0.

The logic and the "scientific method" involved in the proof, with notes, are as follows:

1. An observation is made of many black ravens. (Since this is an observation, it is fundamentally important to logical positivism/ empiricism.)

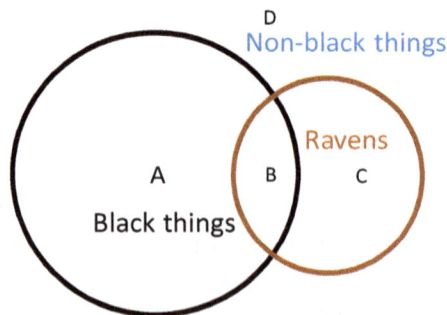

Figure 4.3.   Raven paradox. Circle A represent all black things, and everything beyond the circle A is non-black things. Circle C represents ravens. Overlapping area B of circles A and C represents black ravens.

2. Based on this observation, hypothesis "H", "all ravens are black", is proposed. (This would be *one of the potential conclusion*s drawn by the inductive inference but is presented with a form of deductive logic. Recalling the syllogism in Section 4.2, we have H = "all F's are G" to be proven, where F is for Fact and G for Generalization. Note that the failure of proof is eventually attributed to the inductive inference.)
3. Using H, a secondary hypothesis "H*" is derived where H* is "non-black bodies are non-raven". (This results from the syllogism H* is equivalent to H, no matter whether H is *actually* true or not.)
4. A white swan is found, and it is true that a swan is non-black non-raven. (Therefore, this observation is used as evidence supporting the secondary hypothesis, H*.)
5. Many non-black and non-raven things that are observed are used as evidence for this secondary hypothesis. (H* is supported by a lot of observational evidence and is therefore justified.)
6. Further observation searches for black things and finds ravens among them. (This observation confirms the hypothesis "H".)
7. The hypothesis is proved deductively. (Therefore, one could conclude that "H is true".)
8. With even more observations, non-black ravens are found. (C ≠ 0. H is disproved or falsified. Falsification is discussed more in Section 4.5.)

Let us review what exactly has been proved or not proved from the view of scientists:

Step 1: The observation of black ravens proves that B ≠ 0.

Step 2: Assume that C=0, but it is yet to be proven and is based on the so-called instantial model, which is a form of inductive logic but is not considered scientific reasoning as discussed further in Section 7.3. It is based on the observation of a few samples of F's being G and is used to infer that all F's are G in deductive logic.

Step 3: This is the inversion of the hypothesis. One can use the "If___, then ___" form discussed in Section 4.2. Given that all ravens (*P*) are black (*Q*), if a body is non-black (not *Q*), then the body is non-raven (not *P*). This is in the valid *form* of "*Q* is not true, therefore *P* is not true", like case #3 in Section 4.2. On the surface, this is a valid step.

Steps 4: The observation of the white swans proves that $D \neq 0$, but bears no proof to the hypothesis that $C = 0$.

Step 5: The observations of other non-black non-raven items do not add any new proof to step 4. Again, they prove that $D \neq 0$, but bear no proof to the hypothesis that $C = 0$. The problem is that it does not check "every" non-black thing which is needed to prove H* but was not done. If it is done, non-black ravens should have been found among them.

Step 6: The observation of ravens in black things proves that $B \neq 0$ and is redundant to step 1. It bears no new information.

Step 7: This makes the conclusion that $C = 0$, but no evidence is provided showing $C = 0$ in any proof above. Steps 1, 4, 5, and 6 do not prove $C = 0$. Therefore, the hypothesis has not been proven and the conclusion is drawn without evidence.

Step 8: The concept of falsification is commonly used in science that will be discussed in greater detail in the next section and Chapter 9.

This paradox shows that the logic developed by logicians can be flawed and fallible in certain situations. There is a flaw in step 3. It is true that both H and H* are equivalent forms in terms of deductive logic. However, what is accounted as evidence is different. The proofs given in the next steps are focused on supporting evidence of H*, but no proof has been shown on H itself. Eventually, evidence for H* shown in steps 4 and 5 does not directly contribute to H, i.e., they are *irrelevant* to the hypothesis and, hence, are fundamentally flawed. In fact, H* requires examining "all non-black bodies" which is a harder hypothesis than H to prove and is impossible to accomplish. H can be proved by a complete examination of "all" non-black bodies that prove H* according to the logic but this is not done. This is the second fundamental flaw in the proof. The third fundamental flaw is step 7: The conclusion is drawn without proof showing C=0. This paradox demonstrates clearly that following deductive logic one cannot derive an inductive result.

The idea of constructing deductive logic to derive inductive results is based on the misunderstanding of deductive logic. As we discussed in Section 4.2, deductive logic has to go reductively from more general to more specific, but the raven paradox is a process from specifics to generality.

However, the conclusion that some conventional theories of philosophy of science drew from this failure is more problematic and alarming. The failure of the attempt to deductively prove an inductive result has been attributed to the failure in step 2 that inductive inference itself is fallible. This is a more fundamental mistake, more than the three fundamental logic flaws discussed above. The consequence of this mistake has led (or misled, to be more accurate) philosophy of science to develop "scientific methods" to avoid using inductive logic altogether in science. It is obvious that a logically flawed paradox — drawing a conclusion without direct evidence — does not disprove anything.

The real problem in science is much more complicated than this paradox. If one is unable to identify the flaws in this simple and clean example, with both its method and the language used to describe it, they would be unlikely to conduct any meaningful scientific research. If this paradox is used to argue for the failures of deductive proof of inductive inference and its predictability in many debates, the raven paradox is not very illuminating because of the logical flaws built into the reasoning. Before the flaws are cleaned up, no justifiable conclusion can be drawn from it other than that the paradox is not a real paradox but a flawed logic exercise.

There is a more confusing paradox called the *new riddle of induction*. It is a more widely discussed example in philosophy of science than the paradox of ravens. This "new riddle" contains confusing logic that can be used to test a student's ability and knowledge of logic. It was to build a rigorous deductive theory of confirmation based on inductive logic for logical empiricism using a stricter "form" of deductive logic, but it failed. Therefore, all efforts logical empiricism has made to replace inductive logic with deductive logic have failed. However, to fully appreciate this new riddle and understand the flaws that are included need the forms of scientific reasoning that will be introduced in Chapter 9. Following the formal logic in the traditional theories would not be able to resolve the new riddle. We will discuss it in Section 10.6 as an elective.

Nevertheless, even if the linguistic and logical analysis may indeed reduce the chances of flaws, it is not clear whether the analysis can bring in any new ideas that lead to inventions of new knowledge. Furthermore, scientific problems are seldom yes-or-no or all-or-none questions, although sometimes we do involve conceptual problems that may need a similar analysis. In science, most often the problem concerns "some" conditions or situations. The reasoning should be based more on "conditions" defined by the context of a problem, not according to some abstract forms.

With so many fatal problems, logical positivism and logical empiricism eventually died in the 1970s following Karl Popper's falsification theory and Thomas Kuhn's science revolution theory. Kuhn put it more clearly that science can't be done according to only logic, method, and math. In my opinion, the overall contribution of the ideas of logical empiricism to society is difficult to judge. They definitely raised the awareness of science in the general public, but their theories have also produced more confusion which penetrates deeply into our educational system. Inexperienced scientists or managers of science programs often ask wrong questions or make wrong decisions based on the theory. The negative effects may take more than a few generations to correct as will be discussed in Chapters 7–9.

4.5 Falsificationism

Popper's Theory: Karl Popper (1902–1994) proposed the idea of "conjecture and refutation" (1963, 2014), which was revolutionary in philosophy of science. According to it, science consists of two steps, making a hypothesis and falsifying it, and then making an improved hypothesis and falsifying it. This two-step process repeats forever. His "conjecture" is similar to but less firm than a hypothesis. He argued that science should take more risks and make bolder predictions that can be tested. For example, Einstein made the bold prediction that nothing travels faster than the speed of light. This risk-taking has led to many advancements in modern science. If a theory takes no risk at all because it is compatible with every possible observation, e.g., interpretations of dreams, he argued, it is not science but non-science or pseudoscience. I agree with this basic idea in principle.

Compared with the conventional theory of "hypothesis and justification", the key point is about refutation versus justification. Refutation or falsification emphasizes the negative aspects of a new scientific idea, while the justification or proof, which is also referred to as the verifiability principle, emphasizes the positive aspects. Therefore, Popper's theory is directly against logical positivism in the fundamental aspect. In other words, he believed that justification using the hypothetico-deductive method is false. Popper disbelieved "justification" and thought that confirmation is a myth. According to his theory, a theory cannot be "proved" but only be "disproved" by observation. In philosophy of science, Popper was an induction skeptic who did not believe in the role of inductive logic in science, and hence, Popper agreed with logical positivism in this aspect. He thought that a theory can only be built based on deductive logic.

Demarcation: Popper's theory starts with the "problem of demarcation" that distinguishes science from non/pseudoscience. He uses *falsifiability*, that a theory may "potentially" be proven false, to demarcate it. If a theory is falsifiable, evidence found may either support or be against the prediction of the theory, although the evidence may or may not be convincing enough to completely falsify the theory. If the falsification against a theory is not convincing enough, the theory stands. Otherwise, the falsifier wins; the theory is falsified and does not stand. A theory is unfalsified if it either withstands falsifications or has not been tested. In either case, more falsification tests can still be expected. In his theory, the term "falsifiable" or "unfalsifiable" is different from falsified or unfalsified. An unfalsifiable theory is a theory that cannot be falsified no matter what evidence shows relative to its prediction. According to Popper, this unfalsifiable theory, although withstanding many falsification tests, is nonscience or pseudoscience versus science. Many arguments against Popper's theory were developed on this demarcation.

A theory is not falsifiable, according to Popper, if it is too "flexible" so that no evidence — existing or nonexistent and positive or negative — is able to falsify it because the theory can be argued in a way that is consistent with any evidence. According to Popper, a scientific theory has to make firm (yes or no), or better yet bold, predictions. Furthermore, after a predictions is made, the proposer cannot change. A hypothesis is scientific if and only if it has the "potential" to be refuted, i.e., falsifiable, by some possible observation, regardless of the actual result of the observation.

For example, according to Einstein's general relativity, light bends when passing a massive object. Therefore, the theory predicts that when starlight passes near the Sun, it should bend by a certain amount compared to that when the starlight does not pass near the Sun. This bending may be measurable by telescopes. Given the mass of the Sun, the amount of bending is about twice the accuracy of the telescopes of Einstein's time and, therefore, is unambiguously measurable. Arthur Eddington led an investigation during a solar eclipse in 1919 when they observed a group of stars that would be very near the Sun during the solar eclipse. Without the solar eclipse, these stars are not visible near the Sun because the strong sunlight overwhelms the dim starlight. During the solar eclipse, however, the stars appear bright enough to be observed and the direction of the starlight can be measured. Since the Sun and stars move at different trajectories across the sky, their trajectories and positions in the sky can be accurately determined. The prediction is compared with the observation made during the solar eclipse. The measurements made during this eclipse quantitatively confirmed the effect of Einstein's general relativity. This is an example Popper used to illustrate the falsifiability of a scientific theory because the results, if different from what were measured, could have disproven Einstein's theory. This is the "potential" of refutation or falsification.

In contrast, Popper classified Sigmund Freud's psychology, Karl Marx's view about society and history, and Darwin's theory of evolution as pseudoscience because there are no ways to falsify them. When one shows an observation that is potentially against the prediction of the theory, the theory is flexible enough to be manipulated so that the observation is still consistent with the theory.

The Freudian theory is mostly based on subconscious or unconscious behaviors, such as dreams. There is too much flexibility in interpretation of dreams since a dream can be interpreted in more than two contradicting ways leaving no possibility for refutation or falsification. Similarly, astrology, palm-reading, and fortune-telling are considered to be non-science or pseudoscience at best.

The labeling of Marxism as pseudoscience further exemplifies how demarcation works in Popper's mind. He argues that Marxism predicted, or "conjectured" according to Popper's language, that the communist regime would be first established in an industrial nation such as England or Germany. But the communist regime was first established in Russia in 1917. At the time, Russia was not an industrial country. Therefore, Marx's conjecture was refutable and was refuted, i.e., the prediction of the theory is wrong. However, the followers of Marxism argued that the false prediction was not due to the theory being flawed, but due to the inability of Marx to predict the emergence of Lenin, a charismatic politician and leader of the Russian communist movement. Therefore, Marx's prediction would have been correct if there was no Lenin. This explanation, Popper argued, showed that there is no way to refute Marxism. This deems it as pseudoscience.[3] A problem with Popper's argument is that a scientific theory should be allowed to be modified even if one can argue that it is a different theory, after the modification.

Before we continue, we need note that both Freudian theory and Marxism concern the problem of conduct and not the problem of knowledge. Darwin's theory is the problem of knowledge and worth more discussion. According to his theory, biological evolution is mutation and natural selection. The theory covers an extremely long period of time, spanning a few million years, to potentially include every living or dead thing. A theory of this generality is falsifiable with a large number of potential counterexamples. However, because the uncertainties with the dating and interpretations of these counterexamples are also very large, it is difficult for falsifications to be conclusive. This might have been the reason for Popper to label this theory as pseudoscience in the first place. Additionally, the time scale for "slow evolution" was not defined in Darwin's theory; how long is slow and how short is fast? For example, the Toba catastrophe theory and the extinction of the dinosaurs may have involved some relatively quick changes. The Toba event was a supervolcano eruption that occurred around 75,000 years ago in what is now known as Lake Toba, Indonesia. This event caused a global volcanic

---

[3] One may question whether history, which Marxism was developed to predict, can be classified as a science discipline according to Popper's demarcation. Can anyone develop a science-based general theory of history? Obviously, such a theory would undoubtedly be refutable and could definitely be falsified with counterexamples because there are too many historical events that can be explained in opposite ways.

winter that lasted many years and was followed by a prolonged cooling period. It was believed to have lasted for thousands of years, which is very short in evolution. It is not unreasonable to question, as Popper did, whether Darwin's theory was scientific when it was first proposed. However, after many more decades of investigations, from Darwin to Popper, both the uncertainty and flexibility of the theory of evolution have been reduced substantially, especially when major pieces of evidence can be placed on a reasonable timeline or family tree in relative positions. Therefore, Popper later retracted his labeling of Darwin's evolution theory as non-science. Today, we can confidently state that Darwin's theory is scientific. From this example, we can see that knowledge and science do not seem to be the same thing because the details of Darwin's theory have been continuously added and/or modified. Most people remember only a few catchy slogans and not their science content.

Is Psychology Science or Pseudoscience? In recent decades, many universities have renamed their psychology department as "behavioral science department". This phenomenon is interesting to a scientist. For whatever reason, it seems that adding "science" to the name of a department or discipline appears to lend some creditability to the program. Let us analyze the case of psychology to see whether it is science or pseudoscience. To many people, science is associated with numbers and mathematical equations. If one is playing with numbers and math, they are doing science. Now, the concept of big data is popular. If one had a database that included all information, they would be able to know everything they want to know. Therefore, psychologists are conducting more and more surveys and collecting more and more data. With this data and the analysis, is it true for psychology to be qualified as science if playing with numbers and equations means science?

A few years ago, the news media were filled with all kinds of statistics and surveys, labeling them behavioral science. For example, in one study, the researchers observed behaviors in the elevators of office buildings. There is a certain percentage of people who look at the ceiling or the floor and another percentage of people who are chatting. But, not surprisingly, more people are observed to be using their smartphones. Why would they do what they were doing? Are they too shy or feel too impolite to stare at other people? Or, are smartphones taking over society? No definitive scientific answers to these questions are provided in this study. What can we learn from these statistics about behavioral science? What is the new knowledge? This is not discussed. A similar study was also conducted in bars. Is this science? Because it does not make predictions, it is not falsifiable. According to Popper, it is pseudoscience.

Of course, their colleagues in the same department could argue that behavioral studies are scientific with the research projects that they are conducting. For example, they built a Y-shaped cage in which a monkey can walk freely to each of the three legs of the Y. They put a speaker by each end of the Y and play various types of music or noise at different volumes. The conjecture is that the monkey would spend more time at the speaker that it enjoys more or, at least, hates less. Over time, they may be able to collect a large amount of data on the responses of various animals to different types of music or noise as a function of its loudness. Theories can be developed, and new knowledge can be invented based on the data. Predictions will follow. Everyone can see this qualified as a science project.

Given the two projects in the same department with science designation, how should we classify the department in accordance with Popper's demarcation? Popper classified psychology, history, sociology, economics, etc., as "pseudoscience" because their theories are too flexible and not easily falsifiable. On his list, we noted that most of the pseudoscience disciplines are qualitative in nature and dominated by discovering knowledge-that. Popper's demarcation emphasizes predictions, but knowledge-that has very limited predictability, as discussed in the sunrise and chicken-farmer examples. As more knowledge-that is accumulated, some knowledge-understanding may start emerging in some disciplines. Popper's demarcation indeed has some limitation or ambiguity as shown in the case of behavioral science.

Strictly speaking, if following Popper's definition, all statistics-based science disciplines have problems because any observation can be accepted and understood in these fields just with a different probability. Therefore, according to the demarcation, the theory is unfalsifiable and would be classified as pseudoscience. On the other hand, a theory based on observations cannot withstand falsification because many examples can be used as counterexamples to falsify a theory. The demarcation theory is flawed when applied to many real examples.

Falsification vs Justification: Arguing that the trueness of a scientific theory can never be *proven by observational evidence* is a bold statement that eventually got Popper in trouble. If no theory can be proven true, what is science doing? Can't we measure progress with more positive tests? Should we continue falsifying a theory? Or should we stop if the cost is too high? If we should, how high is too high, many philosophers of science questioned him? I do not see the last two questions as valid in philosophy of science: The objective of science is to invent new knowledge; therefore, time and cost should not be factors in determining truth and knowledge. If we cannot figure out the truth in our lifetime, because of cost or priority concerns, humans will do it later. Therefore, I think that pragmatism, which may be more important for business or engineering problems, should not be involved in scientific judgments, although it could be a factor in the management of science.

When a theory passes more observational tests, is it more likely true or more trustworthy than the one that has not been tested at all? An example that some philosophers of science used against Popper is the selection of designs for building a new bridge. They argue that the design based on existing bridges has been successfully tested and, therefore, presents a smaller risk factor. Conversely, a brand-new design that has not been tested should be riskier. This is a very good example that explains the different perspectives between a philosopher of science and a scientist, as I discuss throughout this book. First, to a scientist, designing a new bridge, in general, is NOT a science problem because it does not involve significant amounts of new knowledge invention. An engineering review committee should be capable of assessing the soundness of the mechanical designs against the expected extreme weather conditions, load, and traffic. No one would build a bridge without a review process, which itself is a falsification process. Second, if the selection is based on whether the design has been tested or not, according to these philosophers, the bridges in the world today would all be the same as the first bridge ever built by humans. Third, building a new bridge mostly concerns artistic impact, financial costs, convenience, and traffic. Therefore, the final selection of a bridge design is more of a political decision versus a science problem. This example contains flaws and is irrelevant to Popper's proposition.

Because science concerns the invention of new knowledge, if an idea in science has not been fully tested, in theory, the very next test could falsify it. Popper's theory has some truth in this respect since there can always be more subsequent tests to a theory; however, he went to an extreme to completely discredit the function of successful justification, which can be seen as an example of the magic power of labeling, as discussed in Section 2.4, when the idea became Poppernian or falsificationism. In science, some understanding can be gained from these successful as well as unsuccessful tests. Often, a theory is modified in order to accommodate the results of tests. Popper's theory overlooked this significant learning process.

One of the most interesting questions to Popper's theory is how do we know if or when we have found truth? According to the proposer of the question, searching for truth is similar to finding the "Holy Grail". The Holy Grail, in this context, is a grail that would glow forever. The problem arises when some grails may glow for a while and then dim. You may hold a glowing grail but cannot be sure whether it will glow forever. The grail goes dim when a falsification becomes successful. You may have to drop it and start searching again. This example describes well one of the two central questions discussed in this book: How do I know I am not wrong? To answer it can be challenging for any scientist. My view on this argument is that searching for the Holy Grail is a bad analogy to describe scientific research because science is seldom a yes-or-no question. Since better theories can result from flawed theories, no one would throw away a grail that glowed for a time, even after finding a more promising one. In this process, people can learn about why the grail goes dim. In other words, scientists may modify a grail to become a different grail or build a new grail based on an out-of-date one, i.e., grails are related. This is the case with Newtonian mechanics. Just because theories of Einstein's relativity and quantum mechanics were invented, scientists would not simply throw away Newton's theory. On the other hand, a scientist never stops questioning the trueness of an existing theory while working with it and never forgets the potential usefulness of a rejected idea. In science, we not only teach students about the existing theories but also the failed ones and the ones that have been improved. We explain the causes of the failure and the reason for the improvement. The question of how we know we have found new knowledge will be discussed in Chapters 9 and 11.

If a theory is falsified, how does one determine if the falsification falsifies the theory as a whole or only a part of a theory? A popular theory in philosophy of science is called "holism", to be further discussed in Section 10.1. It argues that a falsification test cannot be a test of a single hypothesis in isolation but a test of a network of hypotheses. According to holism, falsification of any element within this network may take down the entire theory. This is a valid concern in principle, but in science, most of these possibilities can be identified and isolated. Furthermore, an analysis of the magnitude of different effects can rule out many speculations by holism. The review process for publication is, in essence, a falsification process done by experts. These leading experts would make judgments and recommendations on a problem, or problems, found in the network. Whether the theory is falsified depends on the seriousness, frequency, and magnitude of the problem.

My Criticism to Popper's Idea: The critics of Popper's theory discussed above, although very deliberate, missed a crucial philosophical question: How can existing knowledge falsify new knowledge that is supposed to prove existing knowledge to have flaws? The critics have overlooked an important possibility that a successful falsification based on a network of existing knowledge may be irrelevant because the new knowledge invented may not be consistent with the network of knowledge involved in the test. For example, the Pisa tower free-fall experiment could have been used as evidence for Earth not moving and successfully falsified heliocentrism. The argument did not involve a network of knowledge as expected by the holism — can holists identify where the network of knowledge is false? However, we know the successful falsification itself is false. Galileo would invent the theory of inertia, which was unknown at the time, to better explain the free-fall experiment in relation to Earth's movement. With this new theory, the free-fall experiment can be used as evidence for heliocentrism. Did this make heliocentrism too flexible to be non-science or pseudoscience? If using the same argument that Popper

used to label Marxism as pseudoscience, heliocentrism would have qualified as a pseudoscience because it seems unfalsifiable and because Galileo
modified the original prediction of Copernicus who did not foresee the
free-fall experiment from the Leaning Tower of Pisa.

Another similar example concerns Coulomb's law of 1785. This law
states that electric charges of the same type, i.e., either positive or negative
charges, tend to repel each other and this repelling force is inversely proportional to the square of the distance between the two charges. The law
has been held solidly since then. However, Rutherford's 1911 experiment
indicated that the atomic nuclei are each packed with multiple positive
charges. According to Coulomb's law, these positive charges should have
repelled each other extremely strongly because the distance is extremely
small. Let us imagine that you were a scientist living during the exciting
time when atomic nuclei were first discovered. You realized the conflict
between Coulomb's law and Rutherford's experiment. Were you going to
think Coulomb's law was falsified by the discovery of the atomic nucleus,
in accordance with Popper's theory? Or, were you using Coulomb's law
to falsify the new model of an atom? If you followed Popper's theory, at
least one theory is falsified. Would you throw away the grail of Coulomb's
law as it went dim? Or, if you follow the idea of holism, you will argue
that because there is a web of knowledge involved in Rutherford's experiment, there is no way to determine which of the two theories is right; you
would have either been doing nothing or spent your whole life to learn
about the technical details of each hypothesis and the measurement technique involved. In the 1970s, physicists found that when the distance
between positively charged protons is extremely small, there is another
force, called the strong nuclear force, that takes over and binds the nucleus
together. With the new knowledge, there is no contradiction between
Coulomb's law and Rutherford's experiment. You could potentially be the
scientist who invented the new idea if you did not believe either Popper's
theory or holism.
    From these examples, of heliocentrism and the strong nuclear force,
Popper's theory is seriously flawed because it is falsified according to its
own rule. Note that these two examples are raised by a scientist and not
by philosophers of science. Most criticisms raised by philosophers of science, such as bridge design selection, search for the Holy Grail, holism,

or the referee process, were either less relevant or relatively trivial as discussed in the last two subsections.

Nevertheless, Popper later went to a radical extreme, by thinking confirmation a myth, and has deviated from his original great idea. The radical statement Popper made may have resulted from the contemporary notion of philosophy of science that any theory has to be absolute and exclusive. His original idea, which explicated the importance of falsification in science, was very novel. The requirement of "universality" for a theory of philosophy of science is fundamentally flawed because science itself is seldom an all-or-nothing problem. It is the reason why philosophy of science could not have successfully found a theory that works under most situations. However, Popper's idea definitely brought fresh air to philosophy of science by deeply questioning how we make our decisions and judgments in science. In Chapter 9, I will describe how falsification is used in science.

Questions for Thinking

1. What do you think about sources of knowledge? Is experience or conceptualization more important? Please provide examples.
2. What are deductive logic and inductive logic in your own language? Please provide an example for each.
3. Can you find other possible logic in addition to deductive and inductive logic? What logic is Alvarez's hypothesis based on? Luis and his son Walter Alvarez in 1980 proposed that the extinction of dinosaurs was due to an asteroid impact in the Yucatan Peninsula 65 million years ago. Is it valid logic?
4. Do you think science progressing based on justification or falsification? Please provide examples for your argument.
5. Do you think the following are disciplines of science? Economics, medicine, history, archeology, literature, art, music, psychology, hydrology, ecology? Why?
6. Have you done an IQ test? Is it reasonably correct? In your opinion, are IQ tests scientific or pseudoscientific? Why?

Chapter 5

Science Revolutions and Paradigms

5.1 Intellectual Revolutions

In the eyes of the general public, a scientific revolution may be one of the most fascinating things in science, especially to science-loving minds, as it implies that the things we have known before are all wrong. Philosophy of science definitely fueled the enthusiasm when it discussed the "fall of Newton's theory" with the establishment of Einstein's relativity. The new theory in physics resulted in the development of atomic weapons, which are so powerful that they have the potential to destroy the world many times. However, there is some significant misinformation and misrepresentation in the line of reasoning which we will discuss in this chapter.

Science revolutions refer to fundamental changes in scientific thinking within a relatively short period of time. The length of this time is determined by two factors: the information propagation time and the time for the digestion of the information. In ancient times, the speed of information propagation was slow. Today, during the age of the internet, information propagates at near-light speed to every corner of the world. The matter now is not about the time taken for information propagation, but about the time when the right person pays attention to a specific piece of information. I call this the digestion time of an idea within society.

Immanuel Kant (1724–1804) thought of two revolutions that he called "intellectual revolutions". The first one was the transition of techniques developed by Babylonians and Egyptians, regarding mathematical

111

practices such as geometry and arithmetic, to Greece. This transition must have taken a very long time, say a few centuries. The Greeks used these techniques as tools to prove postulations. Kant's assessment was very insightful because this transition would be the start of "quantitative" science in my view. If you lived before that time, you would have a large range of flexibility when speculating or postulating ideas. You might invent many possible ways of explaining a natural phenomenon, and there would be no way to dismiss a possibility if you were a good sophist. Geometry and algebra, on the other hand, could play a very interesting role in eliminating many potentially possible explanations for a natural phenomenon. Numbers could be a very powerful constraint because there are rules, such as those of algebra and geometry, for how numbers can be manipulated. One would find that many philosophical speculations could not survive this new requirement. This is the power of "evaluation". Here, we have chosen the word "evaluation", instead of "quantification", because some of the science disciplines do not yet have quantitative theories but we require any theory at least to provide some constraint in terms of magnitude of each involved process. It cannot be a random speculation. Modern philosophy of science has, in general, overlooked this requirement so that armchair speculations still dominate the field when the requirement of proof is only about the *form* of logic. However, it is worth mentioning that evaluation here does not involve "moral values".

The second intellectual revolution Kant identified was the concept of experimentation invented by Bacon and the laboratory experimental method developed by Galileo. Galileo conducted many experiments in the lab, such as measuring the acceleration of a block or roller moving on a slide. This could have been the beginning of (quantitatively controlled) systematic experimentation, a pillar of modern science. When the experimental results are analyzed with the geometry and algebra acquired from the first revolution, one finds that some phenomena can be described and even predicted quantitatively. One could imagine that with various slope angles, types of surfaces for the slide, and different sizes and shapes of the blocks, Galileo should have been able to measure the Earth's gravitational acceleration relatively accurately for various settings. This is the power of systematic experiments constrained with evaluation. It can isolate an effect from other possibilities and substantially reduce the candidates for an explanation.

I agree with Kant's two intellectual revolutions — mathematical evaluation and systematic experiments. They are the foundations for modern science. Galileo's systematic experiment methods might be easy to appreciate but using mathematics to judge the success or failure of an idea is somewhat peculiar in terms of philosophy of science. This is because mathematics was developed by a different group of people for completely different purposes, such as to count beans or to draw geometric patterns. This issue will be discussed further in Chapters 8 and 9. With these two pillars accepted, in theory, one could guess an additional pillar that may help to stabilize the system. I am adding a third pillar to the system of science by identifying a third intellectual revolution: Newton's conceptualization and modeling methods.[1]

Although Newton (1624–1727) did some experiments (especially in optics) and invented calculus, his main contributions to science, from the viewpoint of philosophy of science, were *conceptualizing* the known observations and experiments with his three laws of motion and the universal law of gravitation. Each of these laws has a mathematical form and connects with the experiments via mathematics, i.e., evaluation. Recall the discussion in Chapter 4 that there are two sources of knowledge: observation and conceptualization. Knowledge-understanding, which connects different pieces of knowledge-that to reflect the truth, is eventually condensed in the form of conceptualization. Conceptualization was firmly developed only after the concept of "law" of nature was introduced to science. With Newton's laws, we often find it more practical to analyze a mechanics problem by drawing diagrams. We analyze the forces that act on an object and determine its motion by evaluating each force with mathematical derivations, instead of actually conducting experiments. This is the power of conceptualization. This type of conceptualization

---

[1] Kant was born just two years before Newton's passing, and he knew Newton's theory very well. However, he spent much time inventing his own laws of mechanics. He actually invented three laws of mechanics (Walktins, 2019). He did not appreciate the fundamental contributions of Newton, possibly because of the level of the digestion of Newtonian mechanics in Germany at the time.

fundamentally changes the role of mathematics in science by forming a connection between observation and a specific theory or idea.

A "law" of nature provides connections among pieces of knowledge-that. It is based on observations but has been tested under a variety of conditions. It unambiguously and robustly reflects the underlying natural truth and hence can be taken as truth in science. Often, natural laws in science have to be quantitative and require no logical (deductive) proof as long as they truthfully and accurately describe the natural processes. A law of nature has to be able to explain different phenomena[2] and the conditions for its applicability are an essential part of the law. A successful challenge to a law of nature often ends with the modification of the conditions of the applicability of the law.

According to the discussion above, new knowledge invention needs to include three basic elements: observation, conceptualization, and evaluation. Without all three elements, a research discipline cannot be qualified as science. The connection between conceptualization and observation is via reasoning or logic. From observation to conceptualization, it is mostly inductive logic as well as a great deal of intellectual syntheses, while from conceptualization to observation, it is mostly deductive logic. Evaluation makes the connections robust and rigid. In this regard, actually, there are two ingredients in Kant's first revolution: the ancient Egyptian/Babylonian contribution of mathematics/geometry and the Greek contribution of reasoning. Reasoning is the basis for making sense of any experiment. It includes the development of systematic experiments, the development of any model that connects mathematics to the experiments, valid application of mathematics and geometry to the experiments, and the interpretations or predictions of the experiments. Note that I do not use "logic demonstrations" or "rationality", but "reasoning", to indicate potentially diverse possibilities and to distinguish from logic. As discussed in Section 4.4, logic has overemphasized *forms*, which I consider a mistake. A different line of reasoning may result in a different conclusion from the same

---

[2] For example, Moore's law that the number of transistors in a dense integrated circuit (IC) doubles about every two years, invented by Gordon Moore in 1965, is not a law of nature. If Kepler's laws describe the motion of only a single planet, they would not be called laws of nature.

experiment. Among the three elements, experiments and conceptualization are specialized in each science discipline. Reasoning, including mathematics, is common in all science disciplines and is a central topic of philosophy of science. We will discuss scientific reasoning extensively in Chapters 7 and 9.

5.2 Science Revolutions

With the three elements of science, a revolutionary science event may be a historical episode that fundamentally influences a particular science discipline and often sets the foundation for a new science discipline. A fundamental change in a discipline or emergence of a new discipline will have enough power to radiate out to other disciplines and hence over time influence the large-scale landscape of science. I would not use the term "revolution" for some great idea that affects science only at a subdiscipline level or below. Great ideas below the subdiscipline level may be referred to as "approaches". The concept of discipline and subdiscipline will be discussed in more detail in Sections 5.5 and 8.2.

Although Copernicus's heliocentric theory has been discussed as the beginning of the scientific revolution in many popular books of science and philosophy of science, as I discussed in Section 1.4, it was more impactful to political or religious arenas. The theory's true impact on science was very limited, mostly because it did not provide much new substantial knowledge-understanding. After all, the heliocentric theory is a very narrowly focused topic in science and has not evolved into a scientific field beyond a subdiscipline level of science. Copernicus's most significant contribution to philosophy of science may have been raising awareness of the requirements for a "systematism of model development". I recall that his argument for the need for his new model was that the improvements of the original geocentric model could contradict the assumptions of the original model and/or its previous improvements. Although each of these modifications appeared to improve the model's predictions, the weaknesses of the original model were never removed. This is one of the major problems currently in science and engineering practices. For example, when a model is challenged, people often do not defend it with the understanding of how the model correctly describes the process but respond with "it works".

The true knowledge-understanding of planetary motions did not come until Isaac Newton's (1642–1727) universal law of gravitation and the three laws of motion, which I consider to be the first revolutionary event that forms the modern scientific revolution. Newton's theory was built upon Galileo's (1564–1642) development, such as the measurement of gravity and the concept of inertia of motion, and Keppler's (1571–1630) laws of planetary motion and mathematical development of the conic function, i.e., understanding of eccentricity. The impact of Newton's theory is most profound in human history. It not only provides the physical understanding of Earth's gravity but also of planetary motions and the trajectory of a cannonball, under the same framework with mathematical presentations that contain only a single constant: the universal gravitational constant. This constant is supposed to be precise everywhere in the universe — even in other galaxies! All predictions become quantitative; isn't it amazing? It provides the knowledge-understanding of Galileo's free-fall experiment and, at the same time, of Copernicus's theory and explains why we can confidently predict that the Sun will rise again tomorrow morning.[3] However, as I explained in Section 1.4, it also questions the absolute trueness of Copernicus's theory. Another example that is often cited in philosophy of science is using the theory to explain the lunar tides. It later became clearer that the theory could be applied to many more places, such as the designs of buildings or bridges. Almost everything, moving or not moving (when a building or bridge moves, it is collapsing), can be predictable. Newtonian mechanics is the foundation of science for studying things that are visible and not alive. Even for living things or things that are invisible, many concepts from Newton's theory are still widely used.

Another substantial science revolution is James Maxwell's (1831–1879) theory of electromagnetism. Before his new theory, electricity and magnetism were two different fields of research. Faraday's law and Ampere's law, both of which are purely empirical laws, indicated that there are some connections between the two fields. Maxwell introduced

---

[3] As a side note, Hume's problem of induction that challenges our ability to predict whether the Sun will rise tomorrow demonstrates that he did not understand the new knowledge in natural philosophy at his time. He lived from 1711–1776 well after the publication of Newton's *Mathematical Principles of Natural Philosophy* in 1687.

the displacement current to complete a unified theory of electromagnetism by combining Faraday's law and Ampere's law with Coulomb's law. Furthermore, the resulting Maxwell's equations predict that electromagnetic fields can propagate as waves; light is one of them. This revolution is similar to Newton's theory by being based on observation, conceptualization, and evaluation. The impact of this new theory is equally enormous, providing a paradigm for multiple science disciplines. Electricity, optics, radio science, and radio technologies are all based on the theory and are now essential to our modern lives.[4]

Similarly, one may think that every modern scientific discipline that has attracted large-scale scientific research activities was the result of a scientific revolution; the influence depends on the size and intensity of the field. However, conventionally, "science revolution" in philosophy of science refers to events that occurred in the 16th to 17th century when a few more important episodes of the scientific revolution took place.

For example, Antonie Leeuwenhoek's (1632–1723) discovery of microbial life in the 1670s may be considered the starting point of a scientific revolution in biology and has developed into the modern science discipline of microbiology. Leeuwenhoek sold clothing and he used a microscope to inspect the quality of cloth fibers. He found a way to make very fine lenses with large amplification. When he used them to make observations, he saw many things that the naked eye could not see, especially many types of small creatures that we now know as bacteria and microbes. These observations at first were only knowledge-that, without understanding. The knowledge-understanding of these observations was developed a hundred years later by Louis Pasteur (1822–1895), over a hundred years later. We now know that not

---

[4]The field I have been conducting my scientific research in over the past three decades, space physics, combines Maxwell's theory, Newton's theory, and many other theories, such as those of thermodynamics, atomic physics, special relativity, and chemistry. One may consider that the formation of the discipline of space physics resulted from a revolution of a smaller scale than that initiated by Newton's and Maxwell's works. Space physics became a science discipline when scientists put pieces from different disciplines together in order to understand the processes governing the vast areas of space and on the Sun. Since the Sun is a star, the knowledge can be used to understand many fundamental processes in the universe. One of the field's founding fathers, Hannes Alfvén (1908–1995), was a Nobel laureate (1970).

every type of micro-creature causes illness and some of them may help cure diseases or keep people healthy.

Andreas Vesalius and William Harvey (1628) discovered blood circulation and the role of the heart; they made a major contribution to our understanding of our body. Dmitri Ivanovich Mendeleev (1834–1907) invented the periodic table of elements in 1869; the new' understanding built a paradigm between chemistry and physics. Darwin's theory of mutation and natural selection and Mendel's experiment of heredity were revolutionary; each set the foundation for a few science disciplines although not as quantitative as Newton's and Maxwell's theories.

Einstein's theories of relativity and Planck's quantum theory each identified a limit in Newton's theory. We discussed Planck's revolution in Section 1.5 and will discuss Einstein's in detail in Section 6.4; both motivated the studies of philosophy of science. However, neither has overthrown Newton's theory but each established a science discipline in physics.

5.3 Science Paradigms

The concept of a science paradigm was introduced by Thomas Kuhn in his book *Structure of Science Revolution* (Kuhn, 1962). This book recorded Kuhn's insightful observation of scientific developments without invoking contemporary theories of philosophy of science. This approach by itself was a revolution in philosophy of science: Do not conceptualize anything or make a hypothesis first but observe what has happened and what is still happening. He fundamentally changed the subjects of discussion so that many philosophers of science referred him to a historian, as opposed to a philosopher. I think this was a fundamental mistake by these philosophers of science because he started a new paradigm in philosophy of science. Kuhn had solid training in physics and was an entry-level scientist before he changed his career to philosophy of science. Many of his observations and discussions correctly describe the science processes that scientists experience every day. These are quite different from the logical analyses used to debate about philosophy of science by many logicians and linguists as described in Chapter 4.

After the publication of Kuhn's book, due to its focus on the science revolution, the concept of the paradigm became popular — but was misunderstood by many philosophers of science. This idea can be intriguing to the general public, and especially to ambitious scientists who want to invent a new theory and to create a new paradigm that overthrows the old one. The general public has a good appetite for news and theories of scientific revolutions. However, what is a paradigm?

While debates regarding the word "paradigm" have been sparked by Kuhn, the original meaning of the Greek term is *exemplar*. When you are learning mathematics or a science subject, the textbook and professor will first introduce a concept (a law or theory) and then follow it with an example or two showing how to use it to solve a problem. For example, the free-fall theory can be used to calculate the time lapse and speed of a ball falling to the ground from the 3rd floor of a building. When doing homework, students typically look at the problem and compare it to the examples given in the textbook to figure out the intrinsic similarities between the two. One first solves simple problems followed by more complicated ones. This is to use analogy as a reason to derive knowledge and is an important ability for scientists. With practice, it can solve problems faster and more accurately. This way of learning is the original meaning of paradigm. Since Aristotle, it has been known that reasoning based on analogy may sometimes fail but this method is still used widely in education. However, we seldom use the term "paradigm" to describe this way of learning. While some philosophers argue that a paradigm means exemplar, they did not appreciate the evolution of the fundamental meaning of this term. It now means something quite different from that before Kuhn.

In his book, Kuhn clearly explained that in science "a paradigm is rarely on object replication". The concept may be more like common laws based on which people make judicial decisions on various cases. In the *American Heritage Dictionary*, a paradigm is defined as "a set of assumptions, concepts, values, and practices that constitutes a way of viewing reality", and also, "since [the] 1960s [it] has been used in science to refer to a theoretical framework". In learning grammar, a paradigm still means exemplar in the traditional meaning. In science, a paradigm is a theoretical framework that is, in some cases, based on a few fundamental laws or principles with a network of secondary theories and models. For example, tectonic drift theory or Darwin's theory of evolution each is a paradigm and started as a conjecture. These theories and models are supported by a collection of tightly connected observational facts. As explained in Section 4.5, the connection between theory and observation is facilitated by interpretations that involve a theory or paradigm. A paradigm is able and required to unify disparate phenomena. Making connections or unification of various observations is essential to developing knowledge-understanding.

The first mischaracterization by many philosophers concerning the science paradigm is that scientists who share the same paradigm share the same "value". This invokes *value judgment* in defining a paradigm. Scientists concern mostly with the truth or falsity of a theory or idea; *value*, especially moral value, is not among the considerations unless value concerns only simplicity, consistency, and plausibility as defined by Kuhn. In this latter case, the three features of value may be common in competing paradigms as will be discussed in the example in Section 6.5. Therefore, these features are not sufficient for scientists to decide which paradigm to believe. Furthermore, this value system does not change after a science revolution. Kuhn thought that scientists believe in a paradigm based on faith and peer pressure as opposed to rationality and evidence. This was a flawed interpretation of his observations. He underestimated the urge that every scientist wants to propose their own paradigm. Scientists constantly challenge a paradigm. The cause of the coexistence of multiple paradigms is the incompleteness of information. A science revolution is often triggered by a new critical piece of information. A scientist in general has no problem abandoning a paradigm if it is proven false. They actually could be among the first people to challenge the paradigm they work under.

A new paradigm may start as a conjecture. Because of its ability to make connections among disparate observations and, just as any new knowledge-understanding, its backward compatibility with existing theories, i.e., to be able to explain everything existing theories can explain while not producing new anomalies, it wins increasing support from scientists. It solidifies its position and becomes increasingly concrete over time when the paradigm has been successfully applied to different problems. Note that in this process, falsification attempts to the new paradigm could be intense because existing interpretations may be supported by scientists who believe the existing interpretations or paradigm. Most philosophers may be unaware of these attempts. Therefore, each successful new and different application should not be taken lightly or naturally. If our knowledge is a network, connections may first start with a few strings. The intersections of the strings provide some stability to the network. Here, I emphasize that the application of a paradigm to a very different problem has a strong stabilization effect on the network of knowledge. As the paradigm grows, the connections become better and better established with more and more supporting evidence while techniques and mathematical treatments further develop, mature, or even become standardized. If some predictions are *quantitatively* verified, the possibility for alternatives becomes smaller. As our knowledge further improves, the created network of information and knowledge can evolve into more supportive, foundational, and concrete structures with increasing strengths comparable to tents, cabins, concrete buildings, and eventually bomb shelters. Such was the case for theories like Euclidian geometry and Newtonian mechanics.

This is what a paradigm is about. I recognized it rather late in my career after an unsuccessful attempt to challenge an existing prevailing theory, an experience which I will describe later in this section. I wish I had learned about the theory and concept of paradigms in my college years, which is one of the main reasons why I created this book and the accompanying course. However, Kuhn's theory has been under ferocious attacks. The extreme example would be the attack by Margret Masterman (1970) who claimed the word "paradigm" in Kuhn's book meant 21 different things. As a scientist, I read Kuhn's book, as quoted at the beginning of this section, and found the usage perfectly clear, although I do not agree with some of his ideas and statements. In fact, I feel that Thomas Kuhn was the first to correctly describe the structure and process of science in philosophy of science. Now, a question is how a single word can mean 21 different things. How often do you see a word in a dictionary with 21 entries? Or, in an analogy, can a single person be identified as 21 different persons? There is a possibility: to someone who does not know the person well. In this case, any change in clothing, hairdo, or makeup would be mistaken as a different person. Unfortunately, many philosophers, especially those who specialize in linguistics or logic, analyzed the 21 meanings and agreed with Masterman's conclusion![5] I found that most of the criticism was not valid, and I conclude that the criticisms of Kuhn's paradigm theory were primarily based on the ignorance of science or personal biases. A scientist would easily understand what the paradigm is as Kuhn described without much ambiguity.

Let us compare Kuhn's theory with Popper's theory. In Popper's theory, science is permanently open to criticisms, including that of its fundamental ideas. But Kuhn argued that science is not always open to criticism of its fundamental ideas, i.e., a paradigm; otherwise, it would create a chaotic environment. Popper thought science is a series of two-step processes of conjecture and refutation going back and forth forever. In this respect, Popper acknowledges that part of science is a social process. Kuhn, on the other hand, reasoned that there are multiple forms, which will be referred to as phases, of science, including normal science and revolutionary science, which are bridged by crisis science. In the phase of normal science, the validity of the paradigm cannot be challenged. The paradigm is challenged and changes during the revolutionary science phase. These phases will be discussed in the next chapter.

According to Popper, the development of science is based on rationality and evidence, the same as positivism. However, Kuhn argued that science does not develop according to rationality and evidence, but it is based on history. The rationality theory of science, such as "rational reconstruction", is irrelevant to reality. In science, an important philosophical question is as follows: How to know one's rationalization is *not* wrong and how to know the interpretation of the evidence is *not* flawed? Rationality or evidence alone does not guarantee that one is correct. I will discuss this issue in Chapters 7 and 9.

I first encountered the paradigm problem when I was doing my Ph.D. thesis. There was a very famous theoretical model describing the physical

---

[5] It is somewhat remarkable to see that the best place to find the criticism of Kuhn's book is in the introductory essay of the 2012 print of the book by Ian Hacking (2012)! This must have been a very unusual arrangement because Kuhn passed away in 1996 and was unable to respond to these criticisms.

process related to a phenomenon observed by satellites reported in my work. In order to interpret my observation, I studied the model but soon found a fundamental problem in its result. The model specifically describes the effect of the magnetic field in the process. Naturally, if there is no magnetic field, the effect associated with the magnetic field should disappear. However, the model predicts the opposite — the effect associated with the magnetic field increases with decreasing magnetic field strength and is the greatest when the magnetic field goes to zero, which is the farthest from the known result without a magnetic field. This is a textbook-level error or flaw. I was very excited by my finding and worked intensely, writing a paper to challenge the model that had ruled the field for 15 years.

After a few months of effort, my Comment paper was submitted. The editor of the journal sent it to the author of the model to referee it, who was very professional and nice to a student like me, even when I was challenging his theory. In the referee's report, he showed more mathematical details, a few examples of similar treatments that have been considered successful, and physical arguments to support the model although none of them could address the problem I identified. However, the referee's report made it clear to me that I was not simply challenging the *result* of his model; I was challenging his formalism and the framework on which his model was based, i.e., the paradigm! At the time, my experience and training were not sufficient to challenge the framework. Although I knew the model had fundamental errors, unless I was able to pinpoint where exactly the flaw was in his model, my challenge to the model's result was speculation that could not pass the referee process. I had no choice but to withdraw my Comment paper. You may imagine the suffocating feeling I experienced — knowing the model to be wrong but being unable to prove it. This is a similar situation when you witness a murder but could not prove the connection between the murderer and the crime in a courtroom. This is the power of a paradigm! It acts as a tent to protect everyone beneath it if one does not challenge the paradigm itself. The model I challenged was based on a paradigm; therefore, the author of the model was able to use the power of the paradigm to defend it. I was too young and not experienced.

A few years later, two leading scientists in my discipline successfully identified the flaw in the model (Southwood and Kivelson, 1995). In other words, they successfully disconnected the model from the paradigm. In retrospect, if I had learned about Kuhn's theory of the paradigm in college or graduate school, I would have known that unless I could sever the connection between the model and the paradigm, I was challenging the paradigm rather than the model. To challenge a paradigm, even if it is not as concrete as Newton's mechanics, requires much more preparation.

In another example that I will describe in Section 6.5, it took an effort of a small group of us for more than 20 years to establish the framework of a paradigm in our discipline, but it would take much longer, maybe a generation, for people to actually adopt the paradigm. As Planck once said, "a new scientific truth does not triumph by convincing its opponents and making them see the light, but rather because its opponents eventually die, and a new generation grows up that is familiar with it."

One of the ideas that got Kuhn into great trouble is that he thought truth or the view of the world is paradigm dependent. Kuhn did not distinguish truth from knowledge-understanding and science. Therefore, his truth and knowledge/science are all objective according to the conventional theory of philosophy — since truth is objective, the view of the world cannot be paradigm dependent. But on the other hand, because the world view of nature of a scientist depends on a paradigm, according to his theory (objective), truth depends on the (subjective) paradigm, an irreconcilable problem. He was ridiculed by philosophers of science for this less important flaw. According to the definition discussed in Section 3.1, knowledge reflects truth and is not truth. Knowledge-understanding and paradigm are closely related and science is knowledge in development; therefore, knowledge and the view of the world are paradigm dependent while the truth is not.

According to Kuhn's theory, science starts with a state of pre-paradigm science of chaotic fact-gathering and speculation (note, with no hypothesis). Then, a great piece of work unifies these scattered pieces of information to form a paradigm. Kuhn assumed that each field is dominated by a single paradigm at a given time. In a process of consensus-forging, which is similar to setting the rules for a ball game, details or model examples are worked out based on the paradigm. This may be true at the grand level. However, most of the time active scientific research is not conducted at this level. At the level where active scientific research is carried out, the idea that a single paradigm dominates is in general not true. It is the existence of multiple paradigms that attracts and motivates research. When a research topic is dominated by a single paradigm, the only job left is "mop-up" when the best scientists move to more challenging subjects.

In the example discussed in Section 6.5, I have been involved in building an alternative paradigm to unify two current paradigms in my science discipline. The two paradigms have coexisted for more than five decades when people seemed to just choose to play ball games with two sets of rules. The information and evidence can be explained by either paradigm if the information is qualitative. The uncertainty in the interpretations of observations is too large to differentiate the two. The coexistence can only be broken by conclusive new information/evidence with substantially reduced uncertainty that is able to differentiate the two paradigms. To differentiate or unify the two paradigms, new evidence needs to be quantitative.

5.4 Why Are Paradigms Important to Know?

To understand and appreciate the existence of a paradigm is important to a scientist. From the example of my experience with the paradigm, a paradigm seems harmful because it protects flawed ideas. But this is not often the case. It is quite common that when a scientist starts a project, they encounter a problem that cannot be explained with the conventional theory. They have a natural tendency to question the conventional explanation itself. As one learns more and more about the conventional theory, they may start questioning the general knowledge the theory is based on as well.

For example, the day before I started writing this chapter, I received an unsolicited email from a young fellow that I do not know. This fellow's training appeared to be in radio and/or optical signal processing. He apparently works on an autonomous vehicle project. The model result of a signal was different from what he observed. Situations like this are very common in scientific research. Of course, a small difference with autonomous vehicles could kill people, so one should take any small difference seriously. This fellow started questioning the concept of the Doppler shift based on which he measured the speed of his vehicle as well as that of others. This is the same method with which a police officer measures the speed of a traveling car. This method has been solidly derived in theory and is widely used in various applications. Therefore, it can be considered a paradigm. Nonetheless, the fellow questioned the method of the Doppler shift measurement. Since the Doppler measurement method is based on Maxwell's theory describing the propagation of electromagnetic waves, it raises suspicions about Maxwell's theory as well. He goes further to challenge Einstein's general relativity and quantum theory as well. One key point of his argument is that the response time limit of human eyes is 10^{-9} seconds. The limit to the response time of human eyes can produce uncertainties and can contribute to errors of timing in the reception of a signal. He then quoted many historical measurements of the speed of light conducted centuries ago and argued that since these measured speeds vary substantially, Einstein did not know the speed of light accurately enough. Therefore, the constant speed of light assumed in Einstein's relativity is not supported by observations, he concludes. The fellow then invented some new concepts and processes and copyrighted the manuscript that was emailed to me. From this email, one may sense his passion, urge for invention, and frustration for not being appreciated. He must have submitted his invention to publications before sending it to me, but must have been unsuccessful.

In this example, the initial problem is a feature in the measurements that the fellow was not able to explain. He then challenges the Doppler shift calculation in order to explain the difference. The Doppler shift itself is considered a paradigm on a specific topic and is one of the fundamental methods used in astrophysics, astronomy, and cosmology to determine the speed of the expansion of the universe and its age, for example. If one wants to overthrow the Doppler shift, they will have to explain why most of the modern theories in astrophysics and cosmology are all wrong or, at least, need corrections, and why the interpretations of the observations made with the Hubble Telescope are mostly wrong. Certainly, this young fellow is unlikely to be prepared for challenging modern astrophysics and cosmology. His attack on Maxwell's theory and Einstein's relativity, which are higher-level paradigms, makes his task much less attainable because he has to explain why these theories have been successful across disciplines or he has to explain why all these science disciplines are wrong. This is a requirement to challenge multiple paradigms. I recall the requirement of "backward compatible" for any new knowledge discussed in Sections 3.1 and 5.3. If he knew about the concept of the paradigm, he would likely have focused on other possible solutions to the problem in his measurement, such as instruments, data processing, and environment. The chance of finding errors in these areas is much greater than proving that multiple paradigms, such as Doppler shift, Maxwell's electromagnetic wave theory, special relativity, general relativity, and quantum theory, are all wrong only in his case. Each of the paradigms has a mountain of supporting evidence; he has to prove that all these pieces of evidence have been wrongly interpreted. He needs to answer the following question: How did all scientists involved in these science programs not realize the fundamental errors? This is a question we will discuss in Section 11.1. To overthrow paradigms like these is not a job that can be done in a few pages of a manuscript unless the paradigms are already in crisis, which will be discussed in the next chapter.

One should not think of this fellow as an exception in science. There have been countless documented attempts in the past, just like this, plus many more attempts that have not been documented and/or published. In fact, a large fraction of the efforts in science involves attempts such as this. The solidification of a paradigm is made by falsification attempts like this, as discussed in Section 4.5. In this example, this fellow does not lack creativity. However, as I discussed in Section 3.5, without knowing "whether I am fooled by myself", much creativity is wasted, resulting in frustration. Additionally, this fellow is rational when collecting evidence and making arguments; he did not present his arguments according to God or any supernatural forces. By every measure according to the conventional theories of philosophy of science, he is doing science. Though, one may argue that he is not rational because he did not know modern physics well enough to follow the rationality that was already established by someone else. But this was what Copernicus did. Copernicus did not follow the rationality of geocentrism. The conventional theories of philosophy of science have problems here: What is the correct rationalization? Who should set it? This case clearly shows that rationalization has a personal dependency. Each scientist or student of science can learn a great deal from failed attempts like this. Because the structures of science are connected, one cannot simply question a paradigm as a whole based on a relatively small and specific problem.

5.5 Hierarchy of Science Paradigms

According to Kuhn's definition, a paradigm needs to satisfy two conditions: (1) attract an enduring group of adherents and (2) be open-ended with plenty of problems for the group to resolve. Condition #1 states that a paradigm has a social attribute. This is important because our definition of knowledge also has a social attribute. It is interesting to note that in this definition, a paradigm is not defined according to theoretical frameworks or scientific ideas, but by the number of practitioners. It does not require the paradigm to be correct or rational! I agree with Kuhn's observation and think it is extremely insightful. My interpretation of this is that, collectively, a science community (over time) is more intelligent than an individual. Scientists who work under the same paradigm all view good reasons to believe in its framework. The only way for Kuhn's theory to explain the diverse research activities is that there are multiple levels of paradigms. At the top are overarching paradigms. In physics, this refers to things like the laws of Newtonian mechanics, Einstein's relativity, quantum theory, electromagnetism, and thermodynamics. Each overarching paradigm covers a part of the *branch* of physical science. At a level lower, such as optics, fluid mechanics, solid-state physics, plasma physics, and astrophysics, each science discipline builds its paradigm with a set of components from the overarching paradigms, combined with neighboring branches of science, such as chemistry, material science, and biology. Each of the physics disciplines has a relatively solid paradigm. New observations, engineering, and other applications continuously raise new questions to inspire science moving forward.

However, most scientific activities do not take place at the level of discipline. At an even lower level, a subdiscipline may still have a paradigm or multiple coexisting paradigms. But a less developed subdiscipline may not have a solid paradigm, which is commonly referred to as approaches in science. All paradigms, at any level, must be systematically consistent with the higher-level paradigms. This is a new requirement that was not included in Kuhn's theory, but it is crucial. If a paradigm at a higher level undergoes a potential change, ones at lower levels will have to be modified accordingly. It is possible that a newly proposed upper-level paradigm cannot accommodate the lower-level ones. In this case, the lower-level paradigms may become resistant to higher-level changes; the proposer of a higher-level change has to explain why the old paradigm works well with these lower-level paradigms. This is why a paradigm change is difficult but, if successful, is more substantial. In general, the paradigm at a higher level is more rigid than its lower-level counterparts because it covers a broader field of science. The situation when a lower-level paradigm becomes incompatible with the upper-level paradigm will be discussed in Section 6.4.

Some scientific activities can be directed to modify and/or solidify the paradigm. For example, space physics, the field of my scientific research, is a discipline under the branch of space science. It started by combining Maxwell's electromagnetics and Newtonian theory. As the field develops, more and more effects and hence ideas from more paradigms are incorporated. For example, we now need to include elements from quantum physics, chemistry, and relativity so that a paradigm is formed with overlapping fields of electromagnetism, fluid mechanics, statistical mechanics, plasma physics, relativity, chemistry, and thermodynamics, as well as many instrumentation techniques and material sciences. We do not challenge the validity of the overarching paradigms as they provide top-level cover for us.

Our discipline consists of four subdisciplines, which are ionospheric physics, magnetospheric physics, heliospheric physics, and space weather. Kuhn thought that one paradigm dominates each research topic. Instead, at the subdiscipline level, there are multiple paradigms that satisfy the two conditions defined by Kuhn. At the fringe of each of the paradigms, the theory is less solidly developed and may be better referred to as a "commonly agreed upon approach", which has a stronger flavor of "belief". At this level, each paradigm/approach has its own line of classical works that set the framework. Most of the time, a paradigm evolves relatively gradually, but changes can occasionally be more rapid, often as the result of new scientific discoveries/understanding or availability of new models/methods/tools. Most active scientific research is carried out at this subdisciplinary level.

The above description is based on the example of space physics, which may be a more advanced science field with a higher requirement of quantitative evaluations than many non-physics disciplines. It is more active than many science disciplines because there is a strong driver from new observations. Without a constant influx of new observations, it is difficult to sustain a high level of national or international activities.

In my view, many of the science fields, such as biology, psychology, and economics, do not have well-developed paradigms yet. Therefore, they will become more and more quantitative fields of science, which is required for paradigm development. We will discuss fields without solid paradigms in Section 6.3.

One of the inconsistencies in Kuhn's theory is that although he correctly defined the paradigm to have multiple levels, he appeared to consider normal science as everything below an overarching top-level paradigm, and revolution occurs only at the top-level paradigm. When a revolution occurs, even at the top level, it is not as violent as many philosophers of science have described or imagined. For example, Newton's theory never "fell", but just reduced its coverage after two new paradigms emerged in physics, i.e., Einstein's relativity and quantum theory. The fields that Newton's theory is still applicable in do not feel the effect of the revolution much. However, because there are paradigms below the top-level paradigms, some changes to lower-level paradigms can also be revolutionary, which occurs more frequently in science. To philosophers of science, who only observe from a distance over time scales of a few hundred years, these changes are not worth discussion, but a scientist who devotes ~40–50 years of their talent may be directly involved in one of such changes. These changes can be exciting and bright spots in one's career and life that are worth celebrating.

Questions for Thinking

1. Name and describe the scientific revolution events in your discipline. What was the critical issue? What were the alternatives?

2. Is there a paradigm in the field you are studying? What is it? Is it quantitative or qualitative?

3. If there is a paradigm in your field, do you find the paradigm useful (to simplify the line of reasoning in research) or harmful (to protect the old and out of date ideas; note that a paradigm itself does not prohibit scientists from thinking of new ideas or new paradigms)? If there is no obvious paradigm, would people like to have one? Are you interested in participating in building one?

Chapter 6

Phases of Science

This chapter is an adaption of Kuhn's science theory based on my understanding of it and my observations of the science process. Although I make some comments on his ideas, I am more supportive of his theory than of other conventional theories. Kuhn referred to the science process as the "structure" of the science revolution; I prefer to call it "phases of science" (in time). It may be better viewed as dynamic processes. In Chapter 8, I will describe other aspects of the *structure of science*.

6.1 Normal Science

The concept of normal science, proposed by Kuhn, is equally controversial in philosophy of science. It refers to the research activities occurring within a framework provided by a paradigm. Kuhn envisioned it as a well-organized process during which scientists do not spend much time debating about fundamentals. This is a very insightful observation and abstraction made by Kuhn. Why do we need a paradigm to conduct normal science? One has to realize that scientific research is to push the frontier of human knowledge and to invent new knowledge. At this very frontier, we have only incomplete information, and we are all postulating new ideas to predict what is ahead of us. Each of the propositions in the referee process must contain some good reasons or supporting evidence for its existence. However, in this situation, multiple candidates can coexist as the information is insufficient to decide which one is correct or more likely to be correct. The candidates are each correct only to a certain degree constrained by the available information. Scientists then need to devise experiments or detailed theoretical/mathematical analyses, so that more conclusive evidence can narrow the list of candidates or add more required features to a candidate. In cases like this, no one would know, but each predicts, what the next piece of evidence would show. Before new evidence is available, scientists who take a similar approach would agree on some framework, i.e., forming a paradigm.

A paradigm, as discussed in the last chapter, needs to be either existing knowledge-understanding itself or have a well-established line of classical works to which the paradigm can be directly traced. Furthermore, a paradigm needs to be able to explain a few different weakly related phenomena and not include obvious flaws in its assumptions and reasoning. When being questioned on how to get it right, the answer can be simple — it is based on the paradigm. This can save much time validating and explaining the reasoning. However, any "new result" based on a paradigm remains under scrutiny. But the scrutiny usually does not question the paradigm itself by those who accept the paradigm.

In Kuhn's description, normal science is "mop-up" work needed after a scientific revolution. This is a very unfortunate mischaracterization, especially if the scientific revolutions refer to those by Newton and Einstein versus those at a lower level. However, his theory was immediately attacked from all directions in philosophy of science. What is equally unfortunate is that none of these attacks were from the right direction because most attackers thought normal science was not worth doing! By looking at the several publications Kuhn made as an entry-level scientist, it is fair to say that they are mop-up works — making *incremental* progress using the language of scientists. Therefore, he did not have direct personal experience in building a new paradigm or, at even a lower level, an approach before he moved out of science. Although he correctly defined paradigms to have multiple levels, he could not appreciate the differences in normal science under paradigms of different levels. Kuhn specifically discussed paradigms at two different levels: the grand level and the subdisciplinary level. He gave a clear description of these different levels.

The grand level is the whole of society level, covering many science disciplines and is discussed widely by non-experts and media.

For example, when they heard that I am a physicist, my real estate agent, the contractor for my home repairs, and my dentists all discussed Einstein's relativity with me, especially the effect of a slowdown of the clock. To ordinary people, this is the level of science. Science revolutions take place at this level with names such as Newton, Einstein, Darwin, and Planck; however, these are only a handful of names that ordinary people talk about over a period of a few centuries. Most people would not be able to name the Nobel Prize winners 10 years ago in each branch of science. On the other hand, how many scientists have worked during this period across the entire world? There are great scientists from every country in every generation in every science discipline. They contribute greatly to human knowledge development. On the grand level, however, they are mostly unknown and are not as great as the handful of names everyone talks about. However, they are not doing mop-up work either. One cannot say that sending humans to the Moon and back safely is only mop-up work of Newton's mechanics. And, one cannot say that to design the first atomic bomb or the first nuclear power plant is mop-up work of Einstein's relativity.

Normal science takes the grand ideas from the giants of science to solve real problems and/or to apply these ideas to areas that were not perceived by these giants. One may argue that when Newton proposed his laws of motion, he envisioned airplanes, satellites, and even humans landing on the Moon, though there was no solid evidence to back it up. I can say, for sure, that when Darwin wrote his *Origin of Species,* he did not consider that the extinction of dinosaurs could be caused by an asteroid hitting the Earth. The Alvarez hypothesis is not a simple mop-up work of Darwin's theory. Similarly, Planck did not think about the integrated circuits, computers, and smartphones when he proposed quantizing energy, although computers and smartphones comprise integrated circuits that were developed based on semiconductor theories which are based on quantum theory. How many inventions, made by scientists and engineers, have been included in a single smartphone? Are these all mop-up works, and, if so, of whom? Aren't they each a revolution on some level? On Kuhn's list of participants of the "mop-up" works are Euler, Lagrange, Laplace, Gauss, Hamilton, Hertz, Jacobi, Cavendish, Bernoulli, and d'Alembert, the names that we learn about in physics, mathematics, and engineering textbooks. Since Kuhn defined paradigm at multiple levels, the mop-up jobs at the grand level could still be groundbreaking to establish a paradigm at a lower level. Therefore, Kuhn's unfortunate use of the phrase "mop-up" causes confusion because his discussion of science revolutions and paradigm changes is only at the grand level.

This term "mop-up" resulted in a massive dismissal of the significance of normal sciences by philosophers of science. For example, Popper once suggested that normal science sometimes exists, but it is a bad thing and should not be encouraged. In their view, normal science is not creative and is a waste of life and maybe money as well. The general characterization of normal science as not aiming "to produce major novelties, conceptual or phenomenal" is questionable. This should have resulted from the professional bias in some conventional theories of philosophy of science. To many philosophers, only a few things in science can be counted as novelties based on their knowledge. Is the invention of the microwave oven a novelty? I guess their answer would be "yes, but not a great one". But all the theories microwave ovens are based on have existed for decades. There was nothing substantially novel about the science behind this invention. However, to scientists, engineers, or most ordinary people, an invention such as a microwave oven is a novel achievement of a lifetime. Therefore, there is a misrepresentation and misinterpretation of normal science in Kuhn's theory. Using Kuhn's words, the normal science "reformulated" and "produced more substantial changes in the paradigm" to reconcile the difference between Principia (e.g., Newton's theory) and the continental school (on the European continent) which developed sophisticated treatments of "terrestrial problems" (other than celestial mechanics). Note here that "substantial changes in the paradigm" may refer to revolution at the discipline or subdiscipline level. Most non-experts would have difficulty understanding or appreciating a paradigm of the subdiscipline level. How many people can explain quantitatively the principle of the Doppler shift when they are caught by a police officer for speeding?

Just as a paradigm can be global, covering the entirety of human knowledge or a science discipline, and it can be local, covering only a subdiscipline or a subfield of research, normal science is conducted at different levels. On the grand scale, each of the science revolution events discussed in Section 5.2 may be considered the starting point of a paradigm's formation. In the beginning, the new paradigm would be a loosely connected network. Over years, as the digestion of the paradigm deepens and applications spread, normal science strengthens both the main cores of the paradigm and the connections with other disciplines of knowledge. Some areas may start forming more solid patches. On some occasions, the original paradigm may need modifications to accommodate new phenomena and new conditions. Scientists in this phase of science would be among the founding fathers of the discipline, their names appearing in some textbooks. Gradually, the details are worked out and comparisons with more observations are made.

During the gradual development phase of normal science, disciplines and subdisciplines can be developed under the grand-level paradigms. Each discipline or subdiscipline may form lower-level paradigms. Paradigms on this level may experience more frequent modifications and changes. As I explained in Section 5.5, most active scientific research takes place at the subdiscipline level, rather than at the discipline or branch level. For example, even in the simple case of a satellite orbiting the Earth, because the mass is not evenly distributed on the Earth, a satellite's orbit is tugged to one side which has more mass and the orbit is not fixed but in gradual procession drift. The Moon's gravity will add more effects to it. A science discipline of orbital science studies orbits of various objects under the influences of many bodies of different masses and distances.[1] Producing quantitative information is a major function of normal science. Think about how difficult it must be to land a spacecraft on an asteroid as was done by the NASA NEAR spacecraft!

In Section 5.5, we argued that at the level where most active scientific research is carried out, multiple paradigms may coexist in a field. Scientists that follow one paradigm would not be able to convince ones who believe a different paradigm, in a similar manner to the USA and USSR during the Cold War when no one knew if capitalism or communism was a better economic system. Under this condition, scientists would often choose to decline refereeing manuscripts from an opponent paradigm because they each truly think that the works are based on a wrong

---

[1] We recall that Copernicus's theory is not counted as a science revolution because it did not lead to the formation of a science discipline or even a subdiscipline.

paradigm/approach or a wrong set of assumptions or wrong interpretation of observations but cannot prove it with the incomplete information. Each side knows exactly how the other side would respond to the criticisms. It is a reality that, in a science community, if a paper is refereed, i.e., peer-reviewed, by a scientist who supports an opposite paradigm, the debates between the author and referee could appear to be unproductive since they have fundamentally different beliefs to start with. For example, when I served as Editor for a leading journal on space physics, I did an experiment by sending a manuscript to a referee across the paradigm. It immediately set off a fireball since there was too much passion involved on each side. After the heat dissipated, each side agreed that no one could prove a paradigm or disprove the other paradigm unless new and more conclusive evidence is provided. However, my conclusion from this experiment is slightly different from Kuhn's conclusion that communication between paradigms is unproductive. Actually, such communications may help each side to better understand and update the key points on a technical level. Eventually, each side learned what information is most essential in order to resolve the problem. What I also learned is that communication like this cannot take place productively without an impartial arbitrator and cannot be too frequent.

In order to resolve the differences between two competing paradigms, one needs new information. When a new observation is reported, unless it is conclusive, scientists in each paradigm would interpret it differently and/or modify the details of their prediction, but most likely both would use the new observations as evidence for their own paradigm. After strong conclusive evidence is available and assimilated, some paradigms may need major modifications and scientists may regroup.

6.2 Scientists in Normal Science

Normal science is often described and understood as "puzzle-solving" but it may be better characterized as "problem-solving" if puzzle-solving refers to finding the answer that is known in principle to a problem. In practice, to solve a problem with a potential known answer, of which only specifics are not known, is more for engineering than for science, although there is no clear boundary between the two. Most often, science is to solve a "not-yet-solved" problem that appears to have no solution.

According to Kuhn, "a good normal scientist is committed to the paradigm and does not question it". This is a fundamental misunderstanding of scientists by Kuhn and many conventional theories of philosophy of science. Scientists in normal science do not "submit" or "commit" to the rules of a paradigm. As I described earlier, a paradigm has a line of classical works and contains no obvious flaws. It is true that in normal science, any new theories or hypotheses do not challenge the paradigm that covers the subject. Often, following the paradigm is more convenient to start working on a project and has less chance for flaws than inventing an end-to-end theory. If there is no inconsistency between theory and observation, there is no need for science on the topic. In practice, a straightforward project is carried out by students, graduate students, or postdoctoral fellows. Scientists only provide supervision for such a project. However, when you encounter some instances in which the theory and observation do not seem to be consistent with each other, what are you going to do? You first carefully examine your experiment to every minor detail to see whether it is due to some technical errors. But the inconsistency just refuses to go away. Then, the easiest thing to blame is the theory. Yes! The theory is wrong! Is the paradigm correct and applicable? Or, is the alternative paradigm better? Therefore, a paradigm is under constant falsification attempts.

Most of the challenges to a paradigm would fail and never get published because no fatal flaws are identified in the paradigm. Some of the experience gained in these attempts may be summarized in a few statements of the discussion section of a paper. To philosophers and historians, these attempts never happened. However, in science, these failed attempts play an essential role in human knowledge development and take more of scientists' time. This is what Albert Einstein's "99 percent hard work" is truly about. When a fellow scientist asks a challenging question at the end of a seminar given by a good scientist, the speaker should be able to quickly understand the problem and answer the question concisely. Why could one respond in such a short time? It is because they have thought about and rejected this possibility implied in the question during the failed attempts. If the questioner has been working on a related problem, which is why they could ask a challenging question, very likely they had made unsuccessful attempts on this subject too.

A good scientist understands that one cannot simply attack a specific spot on a paradigm. This is not because scientists in the normal science phase lack creativity or are not critical to possible flaws in the paradigm. It is because if one wants to attack the paradigm, they have to attack the whole paradigm that connects many disciplines or subdisciplines, as I discussed in the example of my own experience in Section 5.3. Every piece of evidence that supports the paradigm can be used to falsify the attack.

Why should we care about this? For example, if you lived during the first half of the 1800s and studied the planet Uranus, you first examined Uranus's position. Soon, you would find its position different from that predicted by Newton's laws of motion and gravitation. You would find a systematic difference between the observed and predicted positions. The easiest thing for you to think would be that Newton's laws might not be accurate. Maybe, the gravitational force is not exactly proportional to $1/r^2$, but $1/r^{2\pm\Delta}$ where r is the distance between the Sun and Uranus and Δ is a small correction parameter. If you make a best-fit to all data, you would be able to determine the value of Δ which might be a function of time and direction. You might even find a small correction in the constant of universal gravitation. You would claim that you corrected Newton's universal law of gravitation! Then, you would spend more time learning about the theory. You would find that many places in the theory do not seem correct or reasonable. You had been told that creativity is paramount in scientific research so you would start inventing your own theory and model to explain the observation. You would be able to convince yourself that you had invented a new theory that could explain near perfectly your observed orbit of Uranus based on a standard scientific technique, e.g., regression.

You would eagerly present your new model to your professor or advisor. If your professor/advisor were an experienced scientist, they would take a look at your new idea and say that "it is very interesting, but it is inconsistent with xx-law and yy-observation". "What?" you would ask. "I never used xx-law and yy-observation; they are irrelevant to my problem!" Your adviser would explain to you the inner connection of the laws and observations, i.e., the paradigm. If you were unlucky and were not surrounded by experienced scientists, you would spend much more time on the problem and would write a paper to publish your idea. The manuscript would be reviewed by one or two expert referees. Immediately, the referees of your paper would be strongly against your theory and demand that you explain the accurate and successful applications of Newton's laws to other similar but different cases, such as the orbital motion of the Earth or Moon. Eventually, someone would show that you had made a mistake when questioning Newton's laws because the systematic difference could be produced by a faint star near Uranus that was first recorded in 1613 by Galileo. That faint star, however, is not very far and actually not even a star. It is a planet, Neptune.

This example shows that if you do not know the existence of paradigms, you may have a tendency to question the theoretical framework. If you know that paradigms are something that cannot be easily questioned because of their internal connections with many diverse observations and phenomena, you would have asked different questions and used your creativity in a different way.

In this case, you could have been the person who discovered the planet Neptune, John Adams in 1843, or Urbain Le Verrier in 1845–1846 who independently determined that Uranus's orbit was affected by Neptune when the two were close to each other around 1821. Or, you could have been Alexis Bouvard who first postulated the eighth planet in the solar system. This example shows that unless you have multiple pieces of concrete evidence that potentially challenge a paradigm, you should first try to find solutions within the paradigm or to look at the problem differently so that the paradigm is not challenged.

When you have no choice but to resolve problems within the paradigm, and when you think hard enough, you can eventually find a creative way to resolve the problem or define the problem differently. In other words, attacking a paradigm appears a big and creative idea. In fact, it is often when one has not investigated the problem deeply and wants to take a shortcut to avoid hard work. To challenge a paradigm is much more difficult and riskier than one can imagine because there are countless intelligent people who have tried to challenge the paradigm unsuccessfully. If the paradigm is shaky or in crisis and you have no other choice but to challenge it in order to solve your problem, you have to be much better prepared for the attack. You should not forget the failed attempts until the problem has been resolved satisfactorily. You should now realize that a creative big idea like a light bulb in one's mind, as presented in popular posters everywhere, could often be a wrong idea. The most difficult job for a scientist is to remove most of the creative big ideas that are wrong, i.e., to address the question of "how to know I am not fooled by myself". The removal is conducted by the individual and the team of scientists and not through public debates. Normal science is then the most productive and efficient phase of science. It is not as uncreative or uninteresting as portrayed in some conventional theories of philosophy of science.

From the discussion so far, you should have had a very clear picture of scientists. Good scientists are intelligent, independent thinkers. They are curious and eager to learn while not blindly believing anything they are told. Different from philosophers, they not only ask or discuss questions but also solve real problems. When people with these characteristics come together and form a community, one would guess that there is nothing much people can all agree on, or scientists, like Socrates, debate endlessly on everything they study until they are completely exhausted and continue debating the very next day. However, this is only partially true. How could science make any progress? The answer is that scientists do agree on something. What this something is will be discussed in Chapter 9. But this something is not a scientific method because the concept of scientific methods itself is debatable, as every scientist tends to invent their own clever method. If everyone uses the same method as suggested by philosophers of science, people, at least a large number of people, should derive the same results and there would be nothing to debate. Communication in normal science can be very productive with high efficiency because it saves much time from armchair speculations which philosophers usually do. The details studied in normal science may later reveal something fundamentally new.

The success and power of science are maintained by a delicate balance between the ordered cooperation and single-mindedness of scientists in normal science, together with the ability of orderly breakdown of a paradigm and reconstruction of a new one in a science revolution, as envisioned by Kuhn. In this whole process, science is driven by an invisible hand: the curiosity of individual scientists.

6.3 Science Disciplines with Paradigms in Development

In many disciplines, quantitative overarching paradigms may still be in development. This is true for fields such as biology, psychology, computer, and, to some degree, medicine. Most active science developments occur in fields where there is no well-established paradigm yet. These studies are more descriptive and will eventually become more quantitative. Many of them now have the term "science" added to the names of their branch or discipline, such as life science, behavioral science, computer science, and medical science. Many of these disciplines do not have paradigms that scientists can safely conduct research under. Furthermore, most of these disciplines or branches do not have designated "laws", especially quantitative ones.

Under such conditions, the research activities, more or less, can be considered preparadigmatic science aiming to develop a paradigm with a focus on fact-gathering and technique development, but also producing many speculations. Information gathered from new activities may be less coherent or less systematic without a paradigm. For example, before Newton's theory of motion, the free-fall experiment, observation of planetary motions, and lunar tidal waves appeared to have been isolated research topics. Scientists during this time would have found it relatively easy to propose a new idea or speculation for each topic, most of which, however, would have lived for only a relatively short period of time before they were substantially modified or rejected.

Similar developments can also occur on the fringe of a relatively solid paradigm. This is a more common situation in normal science when scientists slowly, but firmly, push the envelope of knowledge. In these areas, no one knows for sure or is able to predict the correct way to either extend the existing paradigm or to build a new one. For example, in space physics, the lifetime for most results is less than 10 years. When a major new scientific satellite is launched, a significant part of the understanding regarding the subject would be modified or rewritten. To be a scientist during this phase is exciting because there is the potential of finding a lot of new results, some of which can be called new discoveries or inventions. The next generations of scientists use better instruments, computers, and algorithms to analyze new data with better spatial and temporal resolutions. The results are compared to models of finer precision. Then, the result from the older generation is forgotten. However, when one looks back, certain features may appear again and again, although with the better and better resolution provided by newer technologies. Often, it is these permanent features that eventually lead to a better understanding of the problem.

More often, multiple approaches to a science problem are proposed. When the information collected increases, the field would start a fragmentation process, mostly around relatively common and permanent features from observation. Eventually, few approaches, as opposed to one, dominate the field if the information is still insufficient to differentiate them. Paradigm emergence or shift may take place more often and more easily in this period, often triggered by some major new discoveries, new methods, or new instrumentation. However, to non-experts, these opposite paradigms may look similar. During this period in the research field, one has a better chance to make a major contribution to the discipline or subdiscipline. But at the same time, the chance of establishing a paradigm remains small. One's new idea to explain a new result may be quickly falsified by the next new experimental result. This is a prelude to paradigm formation. In the beginning, a paradigm may not be well based and may be vulnerable to attacks arising from new results. The establishment of a solid paradigm often requires individuals with an extraordinary ability to synthesize information of various types and identify intrinsic coherence or connection. Scientists who know philosophy, i.e., who can see the "forest", as Einstein said, will be better prepared to invent a new paradigm.

Branches, such as neuroscience, biology, and psychology, are still in the stage of collecting and digesting information. People know something about the system they are studying; however, the subjects of these branches are more complicated and involve more potential variables. In the case of psychology, for example, "learning" is based on two principles. First, human learning is basically the same as rats and other animals. Second, learning proceeds by reinforcement — behaviors followed by good (or bad) consequences to be repeated (or not). Some people take these two principles as a paradigm to conduct their research. However, both basic assumptions are quite questionable. They have been derived based on observations or experiments related to basic instincts, conditioned reflex, and survival skills learning. The two principles, although widely accepted by educators, are most likely unrelated to the learning of knowledge-understanding even in elementary school, not to mention high school and college. The first assumption might have been true 5000 years ago — before the invention of written language, but is definitely not true now since we learn most knowledge-understanding, as opposed to knowledge-what and knowledge-how, in school and from books. Rats and other animals do not read books, nor do they go to school or take the SAT or GRE exams; their knowledge is not based on understanding.

The second assumption may be true to some young students or some people but is definitely not true to highly motivated people. To a high degree, it does not apply to most rational adults who learn for reasons other than rewards or consequences. As a counterexample, after a failed bank robbery with consequential punishment for the crime, a bank robber would be less likely to stop planning the next attempt but to plan it more carefully while also preparing for possible harsher punishments. Rewards from parents for good grades, such as in the form of monthly allowances, may be useful for unmotivated students, though it is unlikely that they will become scientists who are driven by curiosity and not solely by material rewards. Nevertheless, the popular theory of human learning is highly questionable from the viewpoint of philosophy of science.

The development of new capabilities, both currently and in the upcoming decades, may lead to the formation of new paradigms in the developing fields. In general, weakly connected paradigms are easier to modify. These are among the most exciting fields for talented young scientists to work and the return can be tremendous, mostly because there are no paradigms to restrict one's creativity. For example, Darwin's evolution theory remains qualitative and sketchy. With more accurate dating technologies and gene theory and measurements, more paradigms below the overarching paradigm may be built. Similarly, in economics, there are two major potential theories to develop and become paradigms: invisible hand, which assumes that social economics can be determined by individual self-interest-driven economic decisions, and demand–supply theories, which assume that without interventions the free market tends to reach an equilibrium of demand and supply. Both theories/assumptions have great potential to become more quantitative. More detailed models can be developed so that the description can be more and more quantitative, especially to handle the dynamic process. From these investigations, better upper-level paradigms may emerge.

6.4 The Process of Science Revolution

General Description: A paradigm shift or the emergence of a new paradigm is possible and can more often take place in less developed disciplines, subdisciplines, or sub-subdisciplines of well-developed subdisciplines. According to Kuhn, most often, science is conducted under a shared theory of choice in normal science though we have to modify his theory by allowing the *coexistence of multiple shared theories*. A shared theory provides a relatively accurate prediction and is consistent with major observations on the subject and well-established theories in neighboring fields, as in normal science. As research proceeds, the failure of solving a problem or explaining a new observation may be first treated as an anomaly. However, the number and severity of anomalies may increase with time and eventually crisis occurs. When a crisis occurs, a large fraction of the scientific community would participate in helping understand its possible root causes and find solutions. New theories would appear in the solutions. During the "crisis science" phase, when scientists debate about different possible paradigms, it would be difficult for everyone to evaluate a new theory. However, science is rarely conducted like the social revolutions described by Kuhn mostly because scientists are in general more open-minded to new ideas. If one does not appreciate a new idea, one may decline to comment on it. Even if the scientist is opinionated toward a new theory, e.g., because it is against one's training and belief, one has to provide a scientifically justifiable reason in order to criticize it.

The new theories could coalesce into a new paradigm with an increased understanding of the problem and more applicable situations. Eventually, a paradigm shift may take place. Kuhn's theory explains the general *process* (not a *structure* as he called it) of scientific revolutions nicely.

As discussed in Sections 6.1 and 6.2, it is true that most of the time science is conducted as normal science that is covered under a paradigm. However, some disciplines do not have a solid or commonly agreed upon paradigm which they are searching for, as discussed in Section 6.3. During the course of normal science, inconsistencies will definitely occur between observations and theories. The explanations to an inconsistency may involve a very large range of uncertainty because there are more than 5, if not 10 or more, potential interpretations or explanations for a single inconsistency, especially if the explanation or interpretation is qualitative. Note here that, according to empiricism, observation should be considered carrying truth. But, in reality, a single observation cannot be used to eliminate 4 out of 5 potential possibilities because of incomplete information. However, observations of different types can help eliminate some possibilities. When connecting diverse kinds of observations that are not directly related to the observation in question, some sort of theory or reasoning has to be invoked. If the theory or reasoning is quantitative and not merely qualitative, it is able to conclusively narrow down the potential candidate possibilities. Therefore, a main function of the mathematics used in science is to reduce the uncertainties in conclusions.

Nonetheless, inconsistencies continue to occur for various causes, more due to new observations. Scientists spend more time reconciling the inconsistencies between the theoretical prediction and the observation or reducing the uncertainties in observations and their interpretations. Up to a certain point, the uncertainties and inconsistencies cannot be further reduced, and human factors become negligibly small. At this point, uncertainty and inconsistencies are a result of the available methods, the intrinsic variability and uncertainty in the quantities measured, and the validity of the theory used to explain or predict the observation. If the uncertainty delivered by observation is greater than the ambiguity/inconsistency with the prediction, the crisis will continue to stay. If the uncertainty can be further reduced, say, by new experiments so that it becomes smaller than the inconsistency, the inconsistency can be unambiguously isolated. The inconsistency is genuine in the paradigm and produces a crisis. Many ideas would be proposed to resolve it. According to Kuhn, science revolution will not occur until the emergence of a new paradigm, which is among the proposed solutions. Then, the difference between the new paradigm and the old paradigm can be resolved during a phase of science called revolutionary science. Scientists may shift to the winning paradigm that convincingly resolves the inconsistency.

Special Relativity: One such example is the so-called relative speed of ether. According to ancient theory, the space above the Earth is filled with *ether*, which is a Greek word that refers to pure fresh air. Therefore, ethers should fill the space between the Earth, the Sun, and distant stars. After Copernicus, Galileo, and Newton, the Earth was known to move around the Sun. Maxwell successfully predicted that electromagnetic perturbations, such as light, can propagate as waves in and relative to ethers. Then, an issue was raised about the relative speed between the Earth and the ethers. As we learned in Section 1.4, the Earth's orbiting speed around the Sun is 30 km/s. In half a year, the velocity change can be 60 km/s relative to the ether. This speed should be measurable according to the Doppler shift and Galilean relativity. Michelson–Morley designed a high-precision experiment to measure the relative speed of ether in 1887. To everyone's surprise, the measured speed is within the limits of the instrumentation and is much slower than the Earth's speed of 30 km/s. The same measurements could be made during different seasons when the ether wind may come from different directions with respect to the Earth's movements. The result remains the same — no appreciable ether wind velocity is measured.

This is one of the famous inconsistencies or anomalies that Kuhn referred to. It is an inconsistency that cannot be reconciled in the Galileo–Newton–Maxwell paradigms. As more scientists started investigating the problem, they confirmed Michelson–Morley's findings. Therefore, the severity of the anomaly increased. It refused to be resolved and grew into what Kuhn called a crisis. If you were a physicist living during this exciting time, what would you conclude? Something is definitely wrong. But what is wrong? Don't you want to join in the effort to resolve it? The crisis can eventually lead to a science revolution and the emergence of a new paradigm. In this particular case, Einstein proposed special relativity in 1905, of which the key assumption is that the speed of light in the vacuum is the maximum speed anything can travel at and does not change with the speed of either the source or observer. This assumption is exactly what Michelson–Morley's experiment showed, but it is inconsistent with Galilean relativity in which the speed of wave propagation depends on the relative speed of the source and/or observer. The correctness of Galilean relativity has been supported by uncountable substantial evidence. From our everyday experience, Galilean relativity tells us that the speed of a traveling car measured from a traveling train equals the speed of the train plus (minus) the speed of the car when the two are traveling in the opposite (same) direction. For example, if the two travel at the same speed in the same direction, they are not moving with respect to each other. Similarly, when a formation of airplanes flies in the sky, to the pilots, all airplanes are not moving relative to each other. According to Galilean relativity, if the source and observer are moving away from each other, the speed of light signal received should equal the speed of light plus the speed of the observer relative to the source and is greater than the speed of light.

Therefore, Einstein was challenging the paradigm of Galilean relativity while being consistent with Michelson–Morley's experiment. Einstein's new theory seems natural to explain the experimental result because this is what the experiment showed, but it may be difficult to imagine or even counterintuitive according to Galilean relativity. Is Einstein's new theory "rational" according to what we learned about philosophy of science? At the time of the event, without the knowledge about the development afterward, according to the conventional theories of philosophy of science, one would conclude that Einstein's theory appeared to be crazy and irrational. Would you believe Einstein's theory and doubt Galilean relativity that the Michelson–Morley experiment was based on? If you were a scientist at the time, you would face this question when deciding which side of the debate you would join. Or, would you propose your own theory? Intuitively, it would not be surprising that one believes Galilean relativity to be correct because one experiences it every day. As we all know now, Einstein's theory was eventually correct; the idea that science is rational is false to someone who does not know what will happen afterward. Kuhn's theory is correct on this assessment.

Quantum Theory: We discussed the historical event of the invention of quantum mechanics in Section 1.5. Now, let us examine the emergence of the quantum theory from the viewpoint of philosophy of science. It began with the observation and theory of blackbody radiation. The theory, Wien's approximation, had an anomaly in the longer wavelengths. The Rayleigh–Jeans theory worked well in the long wavelengths but made the problem more severe because it had an anomaly in the shorter wavelengths. This could be a crisis as described by Kuhn. Planck's law appeared to have unified the two descriptions although it was not based on rationality but a mathematical trick; the crisis seemed to be resolved. If Planck's mathematical expression were the new paradigm, the follow-up works should all be "mop-up" jobs.

But no existing theory of philosophy of science and scientific method could explain why scientists did not stop there and started mop-up. Only the curiosity of a scientist, the most essential ingredient in science development, can explain this. Einstein's postulation of the photon appeared, at first, to be a solution of the photoelectric effect but it opened a new era of scientific revolutions. A new paradigm appeared having eventually increased the understanding of the natural world and more problem-solving power of science. It is much easier for historians and philosophers to say decades later that quantum theory is rational; however, at the time, it was not clear to scientists or anyone. Have you had any experiences or evidence that the energies of such small packages are produced, transferred, or consumed? Is the concept of quanta rational or irrational? Again, at the time, was there any indication that the concepts of quantum theory can potentially be used to invent smartphones? This is what scientists did in "mop-up" work, or something not worth doing, as described in some theories of philosophy of science, after a science revolution.

Solar Wind Theory: An example of a smaller-scale science revolution is the solar wind model proposed by Eugene Parker in 1958, which is one of the founding theories of the science discipline of space physics. Before Parker, it was commonly accepted that there was evaporation from the

Sun, called a solar breeze. The solar breeze is a subsonic continuous solar outgassing. One may imagine it like the vapor from a pot of boiling water as we know the Sun is extremely hot. This idea was a paradigm in the early years of space physics. It can be derived elegantly (and rationally) from physical and mathematical treatments.

Parker's solar wind model insists that the solar wind has to be supersonic, actually hypersonic, with a Mach number of about 10. According to Kuhn's definition, at the top level, both solar breeze and solar wind theories are under the same paradigm: Newton's law combined with mass conservation in fluid mechanics. But in space physics, these two are completely different paradigms because when supersonic solar wind encounters the Earth's magnetic field, a shock will form on the Sun-side of the Earth's magnetosphere and will fundamentally change the interaction of solar gas with the magnetosphere. There is no such shock predicted in the solar breeze paradigm. For a few years, both paradigms coexisted in the community when the leading authors of the two models fought bitterly. However, two years after analyzing the measurements from the Explorer 10 satellite, in 1963, the existence of hypersonic solar wind was confirmed. Parker's theory was proven, and Parker became one of the founding fathers of space physics. Since then, the paradigm has shifted from solar breeze to solar wind which has been one of the covering paradigms for space physics, providing an understanding of the space environment that the Sun affects called the heliosphere. In this new paradigm, solar wind particles and electromagnetic fields carry a greater amount of energy which can account for many more severe space weather effects. Does the solar breeze exist? Most likely, because it is one of the solutions to the problem; but no one cares as it must carry too little energy to be significant.

Kuhn's Science Revolution Theory and My Observations: According to Kuhn's principles for the paradigm theory, a paradigm is "a shared basis for a theory of choice, predictively accurate, consistent with well-established theories in neighboring fields, able to unify disparate phenomena, and fruitful of new ideas and discoveries". A science revolution is a collective action in a science community to change from one paradigm to another. For example, Newton's theory completely took over Galileo's theory, Einstein's relativity took over Galilean relativity, and the gene theory combined with the Darwinian evolution theory produced modern biology. I agree with his line of reasoning. However, more often, a new paradigm emerges creating its own space, such as quantum mechanics emerges at the fringe of Newton's paradigm, to form a separate discipline or subdiscipline with the old paradigm remaining intact as the conventional domain. Similarly, Euclidean geometry remains solid after the invention of non-Euclidean. In these cases, one may consider that the new paradigm and old paradigm coexist; however, each carves out the boundary for their validity.

Although Kuhn's theory of the science revolution phase is interesting, historically, most science revolutions did not have a precursor of crisis, e.g., Darwin, Newton, Mendel, or Copernicus (if his theory is counted as a revolution), at least not at the science community level. It is possible that individual scientists themselves feel the crisis and the need for new theories. Revolutionary science is more often triggered by new discoveries or inventions (although not necessarily immediately), such as the rediscovery of Mendel's experiment of genes, new techniques like radioactive dating methods, or new technologies like the microscope and Hubble Space Telescope, to name just a few. One may argue that Einstein's relativity was "triggered" by the Michelson–Morley experiment and that the quantum theory was triggered by the invention of Planck's new formula describing blackbody radiation, as described in Section 1.5; the crisis is a secondary feature of, not essential to, a science revolution. An essential feature of a science revolution is that a science discipline is formed afterward, which I use to definite a revolutionary science.

Revolutions may have a "non-cumulative" nature, through which we both gain and lose something, according to Kuhn. However, I think that the new paradigm has to be backward compatible with the old one in terms of understanding and prediction capability. During and after a revolution, people critically review our knowledge and start a reconciliation and filtering process that removes out-of-date knowledge while modifying and further developing the new theory. Questions that the old paradigm answered may now become puzzling again, as Kuhn argued, and be a drag for the acceptance of the new theory. The root cause of the differences between the two theories will have to be figured out, as part of the establishment of the new theory.

Does a revolution have to be chaotic? Does communication break down during a science revolution in a manner similar to that in a social revolution? Kuhn thought it does because people have different standards of evidence and arguments in a situation that he called "incommensurability". Scientists do tend to be passionate about the theory they believe. The "incommensurability" is due to interpretations based on different beliefs of the same phenomenon. Kuhn used it to describe the difference between Newton's theory and Einstein's relativity. It is clearly not a correct conceptualization because the two theories converge when the speed of the object is much smaller than the speed of light, i.e., they are commensurable under this condition. Incommensurability may exist at a personal level and on a short time scale during a revolution. However, based on my observation, the communication breakdown exists at a relatively low level and not overwhelmingly at a community level. The science community in general is more excited about hearing a new idea. At the community scale, because human knowledge is cumulative, the acute incommensurability among few individuals will be smoothed out over time.

Although with different rationalizations, scientists are more rational than mobsters or martyrs. No scientist would kill another due to scientific disagreement because the science process is operated under the honors system and scientists have to behave professionally, as will be discussed in Chapter 8. Most scientists would follow the scientific reasoning that we will discuss in Chapter 9. The proposers of the many theories of philosophy of science have overlooked these because they have no in-depth knowledge about a science community and science process. Scientists debate intensely. Sometimes, based on my experience, the debates set a high pressure on people involved. However, it is this extremely high personal pressure that helps science make breakthroughs. This situation, when one has a new idea while many scientists are against it, occurs often because the person cannot convince other people in simple terms. This pushes the person to think on a philosophical level, disregarding all mathematics or technical details, in order to be able to crystalize the difference between the new idea and the old idea in a single statement. Ordinarily, this crystallization may take a long time to do. However, the high personal pressure may make it achievable overnight. I have been involved in many extremely intense debates in my career. Many years after a series of debates, when I talked to a scientist with who I argued, the fellow agreed that it was the most productive moment in his career.

Issues Concerning Scientists During Revolution: As discussed above, both concepts of Einstein's relativity and quantum theory were not rational at the time: Would anyone, who is rational, think that Galilean relativity is wrong? Would anyone who is rational think the energy is not continuous but in small discrete packages? Any theory of philosophy of science that assumes science is rational has to first define what "rational" means and then explain who decides whether something is rational. Any discussion about rationality has to be based on incomplete information at the time! When information is incomplete, there are multiple rationalizations. How can one decide which is correct? Furthermore, when first proposed, Planck's radiation law was not based on a quantum theory and had no indication of "quanta" because it was mostly a mathematical maneuver. He did not provide observational evidence for the existence of quanta. Therefore, the characterization of science as a rational and evidence-driven enterprise is questionable or false. Anyone who wants to use linguistic and logical rules to define and guide scientific processes will definitely fail.

How do you know which paradigm is better? We now know that rationality alone is not sufficient for decision-making at the time of the debate. Many conventional theories of philosophy of science agreed upon the criterion of "increased problem-solving powers". Here are a few examples we have to consider. Copernicus's model was simpler, but the initial predictions were poor compared to the geocentric models at the time. The new paradigms of Einstein's relativity can explain the Michelson–Morley experiment, but it did not prove it when it was proposed; it was based on the result. In the early days, it was not clear whether the new theory could explain everything as precisely as Newton's theory did and could be used to produce nuclear power. Similarly, in the early days of the quantum theory's development, although it could explain the radiation spectrum, no one could have predicted that the theory would be used to invent computers and smartphones. Therefore, it is true that each new paradigm/theory eventually increases the problem-solving power long after the debate. For a scientist at the time of the debate, it is not clear what "problem-solving powers" refer to if one does not know what the problem is.

Quantitative understanding and predictions are eventually the best measures of a paradigm's success so a new paradigm must provide a better and more quantitative understanding of the problem, especially with a causal explanation if there is one. Although Aristotle's doctrines might be considered by some as a solidly established science paradigm, it is not quantitative. We learn from the change from Aristotle to Newton that when a paradigm is more mathematically based with quantitatively accurate predictions, it is much more concrete. This is why Euclidean geometry and algebra have stayed for thousands of years, the longest-standing paradigm.

6.5 Note on Effort of a Paradigm Change in Space Physics* (Elective)

In 1996, Eugene Parker called for a paradigm change in magnetosphere–ionosphere physics involving two subdisciplines in which I conduct my research. Since it covers two subdisciplines, the idea/paradigm is at the disciplinary level. The ionosphere is a region in space very close to the Earth, nominally at 90 kilometer to less than 1 megameter altitude. The magnetosphere is a region above the ionosphere that extends to hundreds of megameters from the Earth and interacts with the solar wind. Space physics studies a very complicated system involving many phenomena while the observations are very limited, mostly from in-situ measurements along the trajectories of a limited number of satellites. Obtaining a global picture of the system during an event is extremely difficult. To conduct controlled experiments is nearly impossible.

There are two general theoretical paradigms or approaches to understanding this large region. One is based on the electric engineering (EE) description. In such a description, the magnetosphere and ionosphere consist of elements in EE, such as dynamo, load, resistors, inductors, and capacitors connected with a complicated current system, and the system is evaluated with voltage, current intensity, resistance, impedance, and circuit resonance frequency. The process can be described with equivalent circuit analysis. The characteristic physical quantities are the electric field and electric current. The other paradigm is based on plasma physics, specifically magnetohydrodynamics (MHD) theory. In the MHD paradigm, the system is treated as a fluid mechanics problem and is described by plasma flow and wave propagation that carry the electromagnetic fields. Mathematically, the MHD paradigm involves mass conservation and energy conservation laws, Maxwell's equations, and Newton's laws as well as photochemistry. There is no dynamo or load. The system is described by various forces and energy conversions with characteristic physical quantities of the magnetic field and plasma flow.

In space, the magnetic field and plasma flow are relatively easy to measure directly. The current may be measured under certain circumstances or can be derived from magnetic field measurements according to Ampere's law if there is a constellation of satellites. The electric field may be measured under certain conditions or can be derived from plasma flow measurements with some approximations. In other words, the current and electric field are often not directly measurable but derived under certain limiting assumptions. Nonetheless, the relationship between the EE and MHD approaches can be compared. Ionospheric physics and magnetospheric physics were each developed according to their own tradition based on a line of classical work. Ionospheric physics is mostly based on the EE approach and magnetospheric physics is based more on MHD. The two paradigms describe the space processes in essentially fundamentally different pictures, especially in terms of causal relations.

When studying how the magnetosphere and ionosphere interact with each other, a scientist has to speak both languages in the two paradigms. The two approaches have coexisted for decades; the two sides do not directly attack each other but do not try to resolve the differences. A scientist would choose and develop details of understanding based on one of the paradigms. When more and more observations become available, communications become more difficult. Eugene Parker analyzed the situation and called for an end to the peace in favor of the MHD paradigm which is used widely in other physics disciplines on similar problems. Vytenis Vasyliūnas (2001) examined Parker's analysis and provided more evidence that supported Parker's conclusion. At the time, I had just become interested in the magnetosphere–ionosphere interaction problem in order to develop a better model for space weather forecasts. Vasyliūnas introduced the concept of the paradigm change to me and the two of us started a 20-year effort to establish a unified MHD paradigm for both the magnetosphere and ionosphere while trying to take down the EE paradigm (e.g., Song and Vasyliūnas, 2014). The theory can be potentially extended to the Sun, to the whole heliosphere, and many subjects in astrophysics.

In the 20 years, our discussion has been more about philosophy than about the details of science. That is to say, it was more about how to get it right while not being fooled by ourselves and focusing less on the technical details that we each could carry out. Although this is still an ongoing project, we have established a systematic framework, i.e., an MHD paradigm, for the magnetosphere–ionosphere interaction and solar physics. The conclusion of our investigation is interesting. Recall that we set out to take down the EE paradigm and replace it with an MHD paradigm. We found that the two paradigms were mathematically equivalent under most conditions, although the physical description may be fundamentally different. The two paradigms become different only in the parameter range for magnetic substorms which are a type of intense space weather phenomenon like tornados in terrestrial weather. This is almost the same relationship as the one shared between Newton's theory and Einstein's relativity. I recall that Newton's theory and Einstein's relativity are equivalent when the speed is much smaller than the speed of light. Therefore, we could not take down the EE paradigm completely but only set a limit to its applicability. To most ionospheric physicists, their approach remains valid on the local processes they are studying. However, in order to develop a global space weather forecast model which includes description and prediction of the magnetic substorms, the model has to be based on the MHD paradigm. Scientifically, the MHD paradigm is superior as it is used in plasma physics and astrophysics and provides more understanding and quantitative predictions. The EE paradigm, on the other hand, is less expensive although it will eventually be replaced by the MHD paradigm. From a viewpoint of philosophy of science, it is more interesting to experience the strong and sometimes furious resistance from the scientists with basic training in the EE paradigm. The future is not as optimistic as Planck's solution — waiting until the believers of the EE paradigm eventually die. Because these scientists are educating some next-genera-tion students/scientists with the out-of-date paradigm and the calculations from the MHD paradigm are still extremely difficult and expensive, we may have to wait a few generations for the old idea to completely die out.

Questions for Thinking

1. Have you done any scientific research? What is the nature of it? Is it normal science, crisis science, or revolutionary science? Please explain.
2. What is the most difficult issue you encountered in the research? Do you have difficulty performing the tasks? Can you understand and explain the significance of what you were doing? Do you know how to use computer software? Do you know how to resolve the inconsistencies in the study?
3. Do you find it exciting, interesting, frustrating, not interesting, boring, or a waste of your time?

Chapter 7

Scientific Methods — Failed Attempts

7.1 Searching for a Scientific Method

Traditional philosophy of science has taken developing a theory of the scientific method as one of its major mission objectives for its usefulness and relevance to science. However, the phrase "scientific method" means different things to the general public. You may have heard the term used in the news and public discussions or debates. For example, one claims that a new drug is developed "based on the scientific method". What they mean is often that the drug was developed following a *protocol* or procedure used by the pharmaceutical industry. This means that it used a double-blind test method, a procedure that will be discussed in detail in Section 10.5. In other fields, there are other "standard procedures". These procedures may be based on some scientific theories, but they are not the "scientific method" as discussed in philosophy of science. The scientific method is not about a procedure, but rather it is supposed to be a recipe, or "algorithm" as Kuhn called, that guides people to derive scientific results or knowledge without mistakes. Students start learning about the scientific method in high school or even earlier. In Chapters 1 and 4, I have demonstrated that there are many possibilities where wrong predictions or conclusions may be derived according to the popular scientific methods discussed in philosophy of science. For example, if we had followed the scientific method while testing Copernicus's heliocentric theory, we would have rejected the theory. Kuhn concluded that there is no algorithm for theory choice in science.

However, as we often hear in general public debates, if one side claims their results are based on the "scientific method", the impression is that the results must be "correct". If the other side of the debate cannot claim the same, then there is no need for further debate. But what if both sides claim to be based on scientific methods? Can scientific methods produce two or more opposing conclusions? Can the scientific method itself be flawed, e.g., as we discussed with the paradox of the ravens? Let us first learn about the classical ideas from the giant thinkers in history.

Aristotle's Ideal: One may attribute the first theory of the scientific method to Aristotle. He thought that experience is an important start to an inquiry, although it may not lead to understanding. He introduced the concepts of inductive and deductive logic, as we discussed in Sections 4.2 and 4.3. However, he had deep suspicion about inductive logic and thought it might lead to fallacy. As a result, Western epistemology emphasizes deductive logic. In contrast, Eastern epistemology emphasizes inductive logic although this could be a result of promotion by the rulers. When a controversy occurs, Eastern philosophers tended to find common grounds between opposite views or played down the importance of differences. Given the large range of possible interpretations, one would be able to find commonalities and reconcile the differences, especially in ancient times when the uncertainty in observations or interpretation of them was larger. On the other hand, Western philosophers tended to emphasize and distinguish these differences so that there was less confusion. This general philosophical difference may explain why Western cultures emphasize *individuality*, but Eastern cultures emphasize *commonality*. This may also explain why China was ruled by an emperor throughout much of history, but Europe was mostly organized as multiple smaller kingdoms.

Galileo's Ideal: Important advancements of the method used in science (note that it is not the scientific method as discussed in philosophy of science) were made by Galileo (1592–1610), as discussed in Section 5.1. He introduced systematic experiments and quantitative evaluation to reasoning. Many conventional theories of philosophy of science have underestimated the power of mathematics and its roles. Science has been fundamentally changed. Mathematics, in theory, is a form of strict deductive logic and, by itself, is exact and unique, without ambiguity and uncertainty. If mathematics can handle a chunk of scientific reasoning quantitatively, we need to only worry about the starting point of mathematics and the interpretations of its result.

Systematic experiments with quantitative control provided by mathematics make evaluating a natural process possible. The quantitative empirical relationship between variables can be derived via inductive logic. When this quantitative inductive logic is combined with deductive logic from Aristotle, a complete cycle, or closure, of scientific logic makes it possible for explanation or prediction. Quantitative inductive logic generalizes the empirical information and deductive logic makes specific predictions rationally and quantitatively. One can see that science has been fundamentally changed. Note that this logic cycle is different from either extreme empiricism or extreme rationalism. It does not need to draw an all-or-none conclusion as in the raven paradox or to predict whether the Sun will rise tomorrow morning. Therefore, a scientist does not need to choose between empiricism and rationalism, as discussed in Section 4.1.

Baconian Method: Francis Bacon proposed a method, which is now called the "Baconian method", in 1620. It is "methodical observation of facts as a means of studying and interpreting natural phenomena". His famous saying is, "If a man will begin with certainties, he shall end in doubts; but if he will be content to begin with doubts, he shall end in certainties." Recall that a scientist starts an investigation with a question rather than a hypothesis. In comparison, the most popular scientific method, e.g., the H-D method, starts with a hypothesis, i.e., "certainties"; one can see how far the conventional theory of philosophy of science has deviated from classical ideas of the scientific method. Bacon further reasoned that "For no one successfully investigates the nature of a thing in the thing itself; the inquiry must be enlarged to things that have more in common with it." This means that, in our language, one needs to make connections among several knowledge-that in order to derive knowledge-understanding, as we emphasized in Chapter 4. He did not suggest that the generalizations of observations should be universal or exclusive. Logical positivism/empiricism is a distortion of Bacon's ideal. It is interesting to compare Bacon with Galileo: Bacon is more of a philosopher and Galileo more a scientist. Galileo emphasized more evaluation in science while Bacon did not. One may appreciate the different perspectives between a philosopher and a scientist as discussed in this book.

Descartes's Method: In contrast, only 20 years later than Bacon, René Descartes proposed his scientific method in 1637–1641. We must analyze his ideal independently and not trust the general label some theories of philosophy of science put on him. According to his method, "The first was never to accept anything for true which I did not clearly know to be such; this is to say, carefully to avoid precipitancy and prejudice, and to comprise nothing more in my judgment than what was presented to my mind so clearly and distinctly as to exclude all ground of doubt." Slow down! The first half of the statement is equivalent to the critical thinking skills all scientists should possess. I agree. But what about the second half? Is this to say that our mind holds the final judgment on everything? How do we know our minds hold everything correctly? Do people have the same judgments? If not, will this result in chaos? For science, this implies that simple experiences may or may not be trustworthy as the experience (not the fact itself) depends on a personal interpretation. There is a partial truth to this notion. It could have worked excellently for Descartes himself, noting that Descartes spoke of himself. However, this may not be necessarily 100% true for everyone, as he was a great scientist and philosopher. Each of us has uncountable experiences, good or bad. These experiences have to be rationalized and have to make sense to oneself. Since each of us has a different way to rationalize experiences, we draw different conclusions from similar experiences. Although this may be acceptable in principle, it seems to be a difficult requirement for everyone to agree with each other. How do I know what makes sense to me also makes sense to others? Nevertheless, this is fundamentally different from empiricism which in principle takes observation as the final judgment.

We all know that observation can sometimes distort the truth and some interpretations can be wrong, e.g., the Earth is flat. Even worse, the difficulty in science is that we are often in a situation where we don't have all the information needed to make a sound judgment. Some information could even be wrong. For example, when a straight stick is in water, it will appear bent. Though we know that the stick is not actually bent by taking it out of the water, rather, this is a visual effect caused by light refraction in the water. But if we do not know about certain processes and effects, such as those of light refraction, can we make a correct conclusion? Furthermore, the stick is not always put in the water by us, the scientists. In some cases, we may only be basing our judgments on an image taken remotely. We may have no knowledge about the existence of a different medium between the observer and the target, such as when interpreting an image of a distant galaxy. As described above, I agree with Descartes on critical thinking being a basic attitude with which a scientist looks at things in science. However, either our pure personal intellectual judgment or pure empirical interpretation may not be reliable. This is the problem that each individual scientist has to resolve but on a case-by-case basis. A universal exclusive rule is doomed to fail.

"The second," Descartes continues, was "to divide each of the difficulties under examination into as many parts as possible, and as might be necessary for its adequate solution." This is what we often refer to as "divide and conquer" and is where the concept of the derivative in calculus came from. I agree.

"The third," Descartes adds, was "to conduct my thoughts in such order that, by commencing with objects the simplest and easiest to know, I might ascend by little and little, and, as it were, step by step, to the knowledge of the more complex; assigning in thought a certain order even to those objects which in their own nature do not stand in a relation of antecedence and sequence." This means to begin with the things that are simplest and easiest to understand and proceed according to their order of importance. It is important to point out that the concept of "ordering" needs some "evaluation". This is now a commonly used method in science. For example, in physics, there may be multiple processes affecting a phenomenon with different levels of importance. We often treat a problem first by finding a zeroth-order (the basic) understanding or solution and then add higher-order effects (effects of smaller influence) to the problem to develop a higher-order (more detailed and sophisticated) understanding. In general, I agree with this step but with an additional comment — it is crucially important in this procedure that the modification to the under-standing/solution from higher-order effects cannot fundamentally change the understanding/solution derived from the lower orders. For example, the first-order understanding cannot change fundamentally the zero-order understanding. Careful analysis has to be done in the cases when this occurs. It could be that the initial ordering of the analysis has problems. One of my colleagues believes that it is important to include every effect first and then figure out the relative importance among them. Also, there are mathematical theories, such as chaos, that predict a small change in a parameter can lead to completely different results. A famous example is the so-called butterfly effect according to which an earthquake can be caused by the wing movement of a butterfly thousand miles away. It is based on a flawed logic in science and will be discussed in Section 7.5.

"At the last, in every case to make enumerations so complete, and reviews so general, that I might be assured that nothing was omitted", Descartes further elaborated. I agree with it in principle. This seems an ideal and very optimistic situation assuming that the problem is satisfac-torily solved. But there is a problem in reality: What if the theory does not agree with observation? Should one trust the theory more or the observa-tion? Based on the first step above, "to comprise nothing more in 'my judgment' than what was presented to *my mind* so clearly and distinctly as to exclude all ground of doubt", my interpretation is that he would use his theory to overrule observation. I rather believe that he was not posed this question and, if he were, he could have a different answer. For example, if he were posed the result of the Michelson–Morley experiment, would he trust the result or the Galilean relativity? Before the invention of Einstein's relativity, there was no "ground of doubt" to Galileo's relativity. Therefore, his method is potentially flawed. However, I agree with Descartes more than most conventional theories of philosophy of science, except for the fact that he relied too heavily on his own intellectual judg-ment. If everyone trusts their judgment and if other scientists have a range of varying judgments, whose judgment is correct?

Newton's Ideal: In his thesis, *Principia Mathematica* (*1687*), Newton multiplied, by many folds, the importance of evaluation developed by Galileo and that of conceptualization. Many empirical relationships, such as the free-fall problem and planetary motions, can be understood and modeled in a single framework with a few overarching laws, each of which has a specific mathematical form. Mathematics can then be used to make and verify predictions. When necessary, new mathematics, e.g., calculus, can be invented as he did.

While the physical laws are more general and the mathematics more powerful, a new problem arises: How to use these laws to solve real problems? When are these laws applicable? This is the problem of how to conceptualize a real problem, one of the most important tasks in modern science.

If I have to rationally construct a scientific method based on the theories discussed above, it might look like this:

1. Starting with questions (from Bacon) — not with an understanding or hypothesis. Being critical of everything you hear and not accepting them unless they can be explained simply and intuitively (from Descartes).
2. Following deductive and/or inductive logic (from Aristotle).
3. Conducting systematic experiments (from Bacon and Galileo).
4. Conceptualizing experiments to derive natural laws (from Newton).
5. Conducting mathematical analyses if there are laws that are relevant and applicable (from Galileo and Newton).
6. Dividing a more complicated problem into smaller and simpler problems and attacking one of them at a time (from Descartes).
7. Starting on simpler problems and then adding complexity (from Descartes).
8. Putting the understanding together and abstracting it in simple logic to see whether anything is missing (from Descartes).

One may find that this list combines different things. For example, items 5 to 8 are problem specific.

Mill's Methods: From the discussion in Chapter 4, in science, generalization or induction logic is where flaws may most likely be produced. John Stuart Mill proposed a system of five methods of induction in his book

A System of logic (1843). This is an interesting set of methods although not really the scientific method as in philosophy of science. It is a direct instruction for conducting certain types of science, especially data analyses. I found the idea useful and present it below.

When analyzing observational results, he suggested first looking for common features. These common features show the necessary condition for a phenomenon. This method is called the *direct method of agreement*. Then, one will have to examine the differences among individual observations concerning the phenomenon. This method is called the *method of difference*. The third method is called the *joint method of agreement and difference*. When comparing the common conditions with the conditions that are not common, one is able to identify the necessary and sufficient conditions for the phenomenon. If there are multiple potential causes to a phenomenon, some of the potential causes can be eliminated if they do not occur in every event. The most likely causes can be on the remaining potential list. This is called the *method of residue*. The last method is called the *method of concomitant variations*. With existing multiple potential causes, if the strength of a phenomenon is correlated with the strength of a potential cause quantitatively, this potential cause is more likely to be the true cause of the phenomenon. Note that this is consistent with the "interpolation", as discussed in inductive logic. Mill's methods provide an interesting guideline as to how scientists determine causal relations. These methods are now commonly used in scientific research and will be further discussed in Sections 9.3 and 10.4.

Scientific Method: The term appeared in 1885 together with a description of the method in F. E. Abbot's *Scientific Theism*:

> "Now all the established truths which are formulated in the multifarious propositions of science have been won by the use of the Scientific Method. This method consists of essentially three distinct steps (1) observation and experiments, (2) hypothesis, (3) verification by fresh observation and experiment."

Note that in this description, science starts with observation and not hypothesis and ends with verification, i.e., no falsification is needed.

Now, the scientific method is not only for scientists to follow but is also used in education and for political and managerial decisions. However, the two audiences have completely different objectives from that of scientists. For scientists, the objective is to invent new knowledge and discover new phenomena. The focus is on how to get it right and how to know one is not being fooled by oneself. To the general public, the purpose of a scientific method is to provide confidence in a result and to make judgments when there are multiple contradicting results, although the general public and politicians tend to make judgments according to the status of the scientists or the reputation of the institutions.[1] When there are multiple scientific ideas, there is an opportunity for them to choose the one that best satisfies their personal agenda.

As science is playing more and more important roles in society, "science" and "scientific method" become words with magical power. Students start learning science and scientific method at younger and younger ages. I still remember my daughter's first science project in the second grade. I thought that it was a good opportunity for my daughter to learn some basic science. We both decided to learn about solubility. We put salt, sugar, and sand each in a container with water until the solution was saturated, and then heated the solution and more solute could be added. As a scientist, I thought this a decent learning experience for my daughter. Then, my daughter made a poster and brought the experiments to the science fair. A father of my daughter's classmate came to look at the poster and chatted with me. The *first* question he asked was "what is the hypothesis of the project?" I was totally shocked by the question. Why should a project have to start with a hypothesis? I was speechless. The father was definitely well educated and might be an educator. He must sincerely believe that scientists start a project with a hypothesis. Although

---

[1] In principle, the status of a scientist or an institution's reputation should carry no weight in making a science judgment. My experience has shown many examples of scientists of high status or from reputable institutions making substantial mistakes or erroneous statements in subjects I am familiar with. However, when hearing an explanation of "knowledge" that I have no expertise in, I more likely tend to believe a scientist with a high reputation and from a more reputable institution. I found my behavior ironic. The cause of this inconsistency may be that I can make sound judgments on matters I know well, as Descartes did, but I have to rely on other parameters to help make judgments on subjects I do not know well.

I was a *bona fide* scientist, with the job title of "Scientist" at a National Center, I must have missed that part of education and he was more updated. In my education to becoming a scientist, no one ever told me that I should start a project with a hypothesis. I never heard about the hypothetico-deductive model. I then started learning about it. That science fair was my first lesson on philosophy of science. However, after learning, the theory appears to have nothing relevant to science, I concluded. Several of my scientist friends also complained to me that when they proposed some scientific ideas to their science programs, the managers only questioned "what is the hypothesis?" and then rejected their ideas according to the popular H-D scientific method. One can see that the popular scientific method has become a negative force in science. Then, I learned about Feynman's analogy that there are ornithologists who study scientists as birds. I feel lucky to have become a scientist before I learned about the scientific method.

Next, I will present the reasoning we commonly use to explain a phenomenon and discuss a few methods that are potentially useful to compare with the theory of science as presented in Chapter 9.

7.2 Reasoning

In science, logic is used to connect scientific arguments. In philosophy of science, logic often emphasizes its "form" and relates to symbolic analysis, as well as analytical and synthetic sentence analysis. However, we will learn from the new riddle of induction, in Section 10.6, that "form" does not guarantee the infallibility of logic analysis. Therefore, I choose to use the term "reasoning" or "scientific reasoning" for discussion on how scientists build their arguments. Furthermore, because science requires evaluation, "reasoning" in science, different from logic discussed in conventional philosophy of science, often requires some quantitative constraint. There are many forms of reasoning. Here is a list of them with a description and discussion.

Deductive Reasoning: This reasoning has been introduced in Section 4.2. Deductive reasoning can go as a chain of deduction. It is linear reasoning with the form from A to B ... to C of formal logic from a general to a

specific idea. Mathematics and Euclidean geometry guarantee the infallibility of this deduction, called natural deduction. In science, we more
often use the "if___, then___" form than the syllogism because "if___" is
the precondition for the trueness of the conclusion. For example, the valid
logic form of mathematics is "if P (an equation) is true, then Q (the solution of the equation) is true" form, or case #1 in the form "if___, then___",
from Section 4.2. Also, it may carry the causal relation for A to be the
cause and C the result. It is one of the fundamental forms of scientific
reasoning. We will further discuss in Section 9.2 the issues that one needs
to be mindful of in scientific research.

Inductive Reasoning: As we have discussed in Section 4.3, inductive
reasoning is a generalization made with limited observations, i.e., from the
specific case(s) of *the same* phenomenon to a more general conclusion. It
is not based on formal, or deductive, logic but on observations. Stereotyping
is the generalization of the same type of subjects and, in principle, is a
form of inductive reasoning. However, it specifically refers to the qualitative form of inductive reasoning. Scientific inductive reasoning is a
restricted form of inductive logic as it requires to be quantitative. Other
than specified, *inductive reasoning*, in contrast to either stereotyping or
inductive logic, in this book, refers to either quantitative inductive reasoning or scientific inductive reasoning. In modern sciences, the primary form
of inductive reasoning is not to use a characteristic or a group of characteristics to qualitatively categorize, or stereotype, a type of subjects. It is
not about whether all ravens are black or not. Scientific reasoning is about
quantitative correlations among various variables, more similar to Mill's
method of concomitant variations.

   For example, let us assume that you have eaten three apples in your
life so far and they were all sweet. Now, you are given the 4th apple.
According to "stereotyping", it should be sweet, but when you eat it, you
find it sour. This is a case of the non-black raven in the raven paradox.
This case shows that qualitative inductive reasoning, or stereotyping, is
fallible and in general is not scientific. If you were following the
conventional theories of philosophy of science, you would conclude that
inductive logic fails. However, the fallacy is not due to inductive but due

to being *qualitative*. When you add the evaluation requirement, you question completely different things because you find additional requirements in order to evaluate the conclusion. When induction is *quantitative*, it becomes correlation analyses of several parameters. You would ask about how to define sweetness and study the sweetness or sourness of various types of apples. This can potentially be called a science. Eventually, you may test the sugar and acid content in each type of apple. In philosophy of science, a commonly used example is to inductively generalize the idea of "all metal is electrically conducting". But in science, one would ask how "electrically conducting" is defined. Then, one has to measure the voltage and current for a piece of metal. In fact, one can derive conductivity or resistivity of the metal sample in a quantitative relationship between the voltage and current in a unit cross-section area and length under various conditions.

New ideas often arise to explain these correlations. All scientific laws, such as gravity acceleration, Faraday's law, ideal gas law, and Ohm's law, were generalizations from observations. The rejection of inductive reasoning has been a fundamental mistake of the conventional theories of philosophy of science. Induction is one of the fundamental forms of scientific reasoning and we will further discuss in Section 9.3 the requirements for quantitative inductive reasoning and other issues that need to be careful in scientific research.

Here, we should note that inductive reasoning is a generalization from a limited number of observations to an unlimited number of cases. However, it does not predict what the next observation would be because the prediction for the next one is from general to specific and hence is deductive. In the example of the sweetness of the apples, the conclusion that "all apples are sweet" is inductive based on stereotyping of three sweet apples. However, to predict the fourth apple to be sweet is deductive based on the assumption that the stereotyping conclusion is correct. For a deductive prediction, the validity of every premise has to go through a careful examination for "soundness". The assumption that "all apples are sweet" is clearly not sound. The failure in the prediction is caused by the deduction based on a flawed assumption. Making a deductive prediction is a separate process that has been confused as inductive inference in many theories of philosophy of science.

When many inductive results are collected, a conceptualization can be performed to synthesize, augment, and generalize at a high level. This is what Newton and Darwin did when proposing their great theories.

Abductive Reasoning: In the case described in deductive reasoning above, the cause is known and we predict the result. The reverse logic is called abductive logic or, in our terminology, abductive reasoning. This is when C (result) is known to be true, while the trueness of A (cause) is not known. If we assume A (cause) is true, following deductive/inductive reasoning, we may be able to positively infer C. Then, the conclusion that A causes C is proven to be true. Can't we? There is a problem, a fundamental problem or flaw in this reasoning! This is the logic known as retroduction or *Affirming the consequent* that is recognized as a logical fallacy. It is not acceptable as scientific reasoning, especially in the presence of multiple possible explanations. The proponents of the abductive reasoning admit that there is a remnant of uncertainty or doubt in a conclusion; they retreat and use terms such as "best available" or "most likely" to justify the reasoning. Unfortunately, many theories of philosophy of science consider it the most important way to gain knowledge in science. Let us take a careful examination of the idea.

Abductive reasoning discussed above can be presented as "if C is true, then A is true". Take mathematics as an example which is the most rigorous type of deductive reasoning. If we solve an equation, say $5x - 15 = -5$, we obtain $x = 2$. According to the deductive logic discussed in Section 4.2, valid logic is "if $5x - 15 = -5$ (P) is true, then $x = 2$ (Q) is true". In abductive logic, because $x = 2$ is known, the form is "if Q is true, then P is true". This is case #2 in the "if___, then___" form. But it is an invalid form! Therefore, abductive reasoning is flawed. The argument flow in abductive reasoning is that we can infer the original equation A from a known solution of $x = 2$. The procedure is to assume that equation A is found. Plugging $x = 2$ into the equation finds that $x = 2$ is the solution. Therefore, the claim is that A is the cause of C. This is a type of logic that we call "circular logic" because one assumes A causing C and proves A causing C. The problem is that, as we all know, there are many (uncountable) equations that have the solution of $x = 2$. The proposed equation is only one of the uncountable possibilities; no scientist would agree that you can find the cause of $x = 2$. Therefore, in science, the conclusion from abduction is not trustworthy at all, no matter whether you call it best available or most likely or not.

This reasoning may be acceptable and used to guide some activities in society and sometimes in certain non-critical science projects, but it is not a form of scientific reasoning because other possibilities can derive the same result. For example, polygraph tests draw conclusions about whether one is lying or not based on a person's biological reactions to questions. However, people of different biological and psychological characteristics may react to the questions differently irrelevant of whether they want to lie or not. Especially, after a failed test, an innocent person may become nervous. They may easily fail repeated tests and be mistakenly labeled as a liar, criminal, or spy. This would be a man-made tragedy based on flawed logic. In the opposite situation, a spy can take training to beat the tests and appear not a liar. Another example that demonstrates the potential flaw of abductive reasoning is the case of the famous proverb discussed in Section 7.5.

In science, it is quite often that people develop a computer simulation code to simulate a very complicated process. In a simulation code, many parameters are adjustable. Then, the scientists play with the parameters until the output of the simulation has some "similarities" to observations. The similarities could be "similarly complicated near-random pictures", for example. However, the scientist could make two claims or conclusions: (1) The simulation successfully describes the physical processes shown in observation and hence the code is working and correct, and (2) the parameters when the simulation and observation look similar are the physical conditions when the observation is made. This is a very strange conclusion: The simulation adds only one piece of information, but it results in two results. In an analogy of mathematics, one uses one equation but has solved two unknowns. This reasoning is flawed because the simulation code itself needs to be tested independently in similar processes to be studied before drawing any conclusion about the process. This is flawed reasoning similar to abductive logic. Unfortunately, currently, there are many such cases in many science fields of scientific research.

In science, we require (1) independent evidence for the cause A, (2) deductive and/or inductive reasoning in each step to reach C, and (3) no other possibility that can reach C. With these requirements, the reasoning is no longer abductive; we actually reconstruct the deductive and/or inductive reasoning from A to C. Therefore, abductive reasoning is deeply flawed and not scientific. Traditional philosophy of science has used it to pollute the general public about science. Many philosophers of science argued that abductive reasoning is the most creative part of science. It is true that in science, abductive reasoning may help a scientist to conceive an idea; it cannot be used as proof of the idea. Practically, scientists spend most of their time filtering out most of their ideas. Without knowing this reality, one does not know what scientists are doing.

Dialectical Reasoning: Dialectic is a style of argument developed in the pre-Socratic period. It is commonly agreed that it was started by Zeno of Elea (495–430 BC). This method is based on contradiction or "reduction to absurdity". Zeno introduced many paradoxes, such as Achilles and the tortoise, the dichotomy, the arrow, and moving rows, just to name the four most famous ones. For example, the dichotomy argues that a stick can be cut in half and the half can be cut in half again. This process can continue forever as the piece that remains becomes smaller and smaller. This concept may be considered the origin of the mathematical concept of the limit and derivative. The remaining part can be represented by a series of 2^{-n} which goes toward the limit of 0 as n increases but cannot be 0. An interesting story is that at a slightly later time, a Chinese sophist, Hui Zi, also debated about the same paradox. He famously defined paradoxically what the largest is and what the smallest is. "The largest thing has nothing beyond it; it is called *the One of largeness*. Similarly, the smallest thing has nothing within it; it is *the One of smallness*." He pointed out that the problem in the dichotomy paradox is caused by the definition of the "end" as the end may be the point that cannot be divided. The paradox is not self-consistent. In modern science, we now know that when a thing is divided small enough, to the atomic level, the thing becomes something else that is fundamentally different from the original thing. Therefore, the division does have a limit in terms of its physical properties. Since dichotomy has been considered a centerpiece of dialectic, dialectic in philosophy is considered an absurd theory as it leads to many paradoxes. It is not considered a scientific method in traditional theories of philosophy of science.

However, in logic, "reduction to absurdity" can be used as indirect proof, i.e., if one cannot prove a premise to be true due to the Popper falsification theory, one may be able to prove that it cannot be wrong. One can prove this by assuming that if the premise is not true, the conclusion is wrong. Therefore, the premise may be correct. A problem in science is that there are multiple possible premises that lead to the same conclusion. Even if one proves one particular premise to be potentially correct, one still does not know what is correct. Therefore, this method cannot be used to prove but only to disprove. Most conventional theories of philosophy of science try to avoid discussing dialectical reasoning altogether. I still remember vividly that once I mentioned dialectical reasoning to a highly respected philosopher of science; his reaction was like I had mentioned the devil. Dialectical reasoning may not follow any rules in philosophical discussions, resulting in most armchair speculations.

In science, however, dialectical reasoning is used restrictively. It can be constructed in different forms for various purposes but often starts with a *reasonable* and *relevant* assumption — "If…." Then, it follows strict deductive and/or inductive reasoning until a *negative* result is derived. In contrast, abductive reasoning is when a positive result is derived. We now still use the same notations as in the abductive reasoning above. In this case, result C is known and we are testing a reasonable/relevant potential cause A. Dialectical reasoning starts with an assumption "if A is true". Then, one follows deductive and/or inductive reasoning to prove C. If the conclusion is wrong or absurd, C is negatively inferred. The conclusion of the dialectical reasoning is that A is false. In other words, the possibility of A being the cause of C can be eliminated. For example, in geocentricism, the Sun rotates around the Earth. *If* the Sun rotates around the Earth once a day as we draw from direct observation, the speed of the Sun would be $2\pi \times 1AU/1day \sim 100,000$ km/second. This is an absurdly high speed that needs explanation by the geocentric model. In this example, we start with "IF…" and follow strict deductive reasoning to derive "Then… is not true". Since it is used for disproof, the falsifier does not need to provide an answer to the problem, but the proposer needs to explain why the theory can work in that situation. If the explanation is consistent with the original theory and with other commonly known facts, the concern is considered addressed.

In this particular form, the first part of the procedure is similar to abductive reasoning, only requiring negative results. The logical form of this reasoning is "if Q (result) is not true, then P (cause) is not true". This is logic #3, discussed in Section 4.2, and it is a valid form. We recall that the logical form for abductive reasoning is "if Q (result) is true, then P (cause) is true", the logic #2. Popper's falsification theory is the same as dialectical logic. The only difference from Popper's idea is that dialectical reasoning has to follow the exact same strict logic as the proof and the starting assumption has to be *reasonable* and *relevant*. In the science process, all possibilities are allowed to rise. However, the reasoning for each of these possibilities is subject to rigorous falsification, i.e., dialectical reasoning. This is a method of elimination by trial and error for all reasonable and relevant possibilities. It is one of the fundamental forms of scientific reasoning. In Section 9.4, we will further discuss issues that need to be handled carefully in scientific research.

Analogy: An analogy accepts logic A (cause) to C (result) in one process and uses it to prove A′ to C′ in a different process where A′ and C′ have some similarities to A and C, respectively. It is a kind of generalization of causal relations and hence it is a form of inductive logic. However, analogy uses characteristics *qualitatively* and the generalization of characterization is applied to *different types of subjects*. In science, since the analogy is used to generalize from one type to another type of subject, it is fallible.

The analogy is a very useful way to *describe* a truth or knowledge, but it *does not prove* the truth and cannot be used to invent knowledge because A and A′ or C and C′ may be different fundamentally or may not be comparable in some key characteristics. For example, conventional theories of philosophy of science have used the analogy of scientists to Russell's chickens and Pavlov's dog in the learning process and the analogy of "searching for the Holy Grail" to searching for truth. Most people would not agree with these analogies since the objects in comparison are fundamentally different. These analogies may be used to describe an idea in philosophy of science but are not acceptable as proof of new knowledge.

Citations: In science, citations can be used to present evidence and to simplify the presentation of reasoning. However, unless the cited work directly contributes to the argument to be established, it suffers the same problem as analogy reasoning. Furthermore, the cited work, although lending support to the new idea, also reduces the degree of *newness* of the new work. It is more often used to defend and less to promote the new work. It is acceptable as scientific reasoning under certain conditions for certain purposes.

Exhaustive Search/Brute Force Method: Although this is more of a problem-solving method, not a type of reasoning, it is sometimes used as reasoning in science discussion. With this method, one tries many possibilities under a specific condition. It can be used to either prove a theory abductively or to disprove one dialectically. It does not provide conclusive proof (disproof) but rather argues that no (at least one) negative case can be found. The raven paradox is always here to haunt this method until knowledge-understanding is developed. It may be applicable to discrete problems. In modern science, this method may be easily implemented with a computer, though it is not scientific reasoning. However, it is acceptable under certain circumstances, especially when the method is used to defend a new work while criticism is based on dialectical reasoning.

Numerical Simulations: There are many numerical simulation models for various purposes based on all kinds of theories. In science, we categorize simulation models as first-principle-based models or non-first-principle models. A first-principle model is based on quantitative natural laws and numerically solves the involved governing equations. This type of model is based on deductive reasoning; however, the numerical methods may introduce uncertainties which will be discussed in Section 9.2. All other numerical models, such as those based on neural network or artificial intelligence theory and those based on empirical data using some sort of interpolation or extrapolation, are non-first-principle models.

These models are widely used in the engineering and financial industry. Some people claim that artificial intelligence has been used successfully in predicting the stock market. This is a dream. Many people claimed the same thing before, such as using derivatives. The reality is that many of these people went bankrupt. The reason is simple: If a theory can truly predict the stock market, everyone would adopt it; but it is impossible that everyone in the stock market makes money. Because the non-first-principle numerical models are not based on knowledge-understanding or scientific understanding, they are not accepted as scientific reasoning.

Pragmatic Argument: This argument is not based on scientific reasoning but on economic or abductive reasoning when one potential answer is found while it was not affordable to search for other possibilities. Because science is seeking truth, cost and time cannot be used as a valid argument to prove knowledge that reflects the truth. The argument is not acceptable as scientific reasoning. But it can sometimes be used to break an investigation into multiple phases or steps.

Occam's Razor: This is also known as the argument of simplicity. When two or more equivalent conclusions are derived by different approaches/ models, the simplest approach/model is "chosen" as correct for the time being. One of the questions is as follows: If we start from the same point and end near the same point, does this imply that the long road with more turns is wrong or not? We note that the longer or shorter route is about good or bad, not necessarily about true or false. Some people invoke aesthetic reasoning to justify this argument. One is able to find some examples, as will be discussed in Sections 7.6 and 9.5, in which simpler models or approaches are later found wrong.

I prefer the argument of probability as the reasoning for Occam's razor. An approach may involve a few assumptions or logical steps; each of the assumptions or steps, especially qualitative ones, may inherently carry certain uncertainty which reduces the probability of the success of the approach. Therefore, the approach/model with the fewest assumptions or steps and with more quantitative control is more likely to reflect the essence of the natural processes. Occam's razor is *considered* as one of the fundamental forms of scientific reasoning. For example, in the geocentric vs. heliocentric debate, we *take* heliocentric as *more* "correct".

In the following, I discuss a few commonly used methods or models.

7.3 Instantial Model

This may be the most used method by most people in our everyday life. Often quotations or interviews in news reports make people think that the whole world is like what the reporter wants people to believe. For example, a news report may show the results that a certain percentage of white or non-white, male or female people support proposal A or proposal B. The watchers may conclude that every issue seems to be decided by race or gender. Or, the report may select a few clips with extreme points, although all sound reasonable, of view on an issue; there seems no possible solution to the issue. The root reasoning for the generalization of the instantial model is *qualitative inductive* logic. Since the generalization is qualitative, it is either an analogy or stereotyping, as discussed in the last section. Logically, the conclusion is based on a few facts or observations to generalize all in that category or even other categories in the case of analogy. We know that reasoning by analogy or stereotyping is fallible and not scientific.

This is logically represented as follows:

"An F is G" accounts as evidence for "All F's are G".

The raven paradox, described in Sections 1.3 and 4.3, is an elaborated version of this logic. In the raven paradox, even if one does not conclude that all ravens are black based on one or a few observations of black ravens, the logic is the same. This logic is called instantial inductive reference in philosophy of science. The user of the logic may be subconscious or may make extra effort to prove the hypothesis, such as looking for ravens in non-black things or in all black things. Logicians have categorized this model as inductive logic and its failure in prediction is attributed to inductive logic. This is a debatable conclusion because they do not distinguish the qualitative (analogy or stereotyping) from the quantitative (correlative) form of inductive reasoning, such as the method of concomitant variations described in Mill's method. The latter is scientific.

This model is used either forward to predict future events or backward to explain or abductively prove the causes for what we are seeing now. Sometimes, we make predictions based on a "guess" and other times, we do it based on causality. As we have learned, predicting future events is deductive and requires knowledge-understanding. The fallacy of generalization models in everyday life may not always be important. This logic is potentially useful in triggering new ideas but is not acceptable as proof of a scientific idea. Therefore, this is not a scientific model. In science, an idea can rarely explain "all" things or "everything". This could be the root problem of the instantial model.

7.4 Hypothetico-Deductive Model

The most cited scientific method in philosophy of science is the so-called "hypothetico-deductive" (H-D) model. The original proposer, Carl Hempel (1905–1997) is one of the founders of philosophy of science. He argued against the idea that science needs first to collect "all facts". According to his idea, without selection or guess, such a science project "could never be carried out, for a collection of *all* the facts would have to await the end of the world, …". I consider this a rather radical argument. Maybe his opponent should have included a modifier "available and potentially relevant" between "all" and "facts". However, he could argue that who decides what is relevant; this sounds like a very poisonous debate that most scientists would not want to engage in. Nevertheless, with such an argument, he established, among fellow philosophers of science, that science should start research with a hypothesis in order to reduce the relevant facts; strangely enough, everyone else seemed to agree with it without consulting with scientists.

Because of the known problem with inductive logic, induction was not included in a scientific method. With deductive logic as required, he derived the H-D model. To some people, the word "hypothesis", by itself, sounds scientific. This model is now taught to pre-college students as part of learning about science. According to Encyclopedia Britannica, the "hypothetico-deductive method, also called H-D method or H-D, procedure for the construction of a scientific theory that will account for the results obtained through direct observation and experimentation and that will, through inference, predict further effects that can then be verified or disproved by empirical evidence derived from other experiments." This definition does not include either the "hypothesis" or "deductive" elements of the method, indicating a deviation from the original definition. The following are some examples that have been used in the philosophy of science discussion. In each example, I add some comments that question the conclusion drawn in the conventional theories.

Hypothesis: "The early bird gets the worm." No observational basis is needed in the H-D model, but an old proverb.

Deduced prediction: Early birds weigh more. This is an interesting but flawed deductive inference.

Test: Statistically, early birds weigh more; conjecture is proved. The problem is that the conclusion may or may not be conclusive because there are more effects on the weight of a bird. The amount of food does not necessarily result in gaining more weight as one may use more energy via more activities. Can we say that all overweighted people eat more? It could be that some birds are genetically more active and more energetic. They get up earlier and are seen as early birds but do not necessarily eat more worms in the early morning. This conclusion is not conclusive.

Hypothesis: "Light is a wave."

Deduced prediction: Diffraction (for a shade with a small hole) or interference (for a shade with two slits) occurs for waves when the image of the hole or slits formed on a screen behind the shade is very different from the hole or the two slits. This is a valid deductive inference but the prediction is based on knowledge-understanding.

Test: When the hole is small enough or the slits are narrow enough, diffraction and interference occur. The hypothesis is confirmed. The conclusion is conclusive but may not be exclusive because one can prove a hypothesis that light also behaves like particles, i.e., photons that carry momentum. This is knowledge-understanding with sufficient information and is still not able to exclude the possibility that light is made of particles.

Hypothesis: The Earth is not moving. This can be based on everyday life experience.

Deduced Prediction: After releasing a ball from the Leaning Tower of Pisa, when the ball is falling, if the Earth is not moving, the ball would hit the ground at a spot just beneath the dropping point. This is a valid deductive inference before Galileo and Newton.

Test: The ball hits the ground right beneath the dropping point. The hypothesis is proved; the Earth is the center of the universe while we see the Sun and Moon rise and set. However, the conclusion is wrong.[2] The conventional theories of philosophy of science do not discuss this failure.

From these examples, the H-D method often cannot arrive at conclusive scientific results and at times it may lead to a questionable or completely wrong conclusion. The method may be interesting for school students to learn about science, but it is not useful for scientists. On the internet, one may find many improved versions of the H-D procedures. Most of these improved versions include the observation and data collection as the starting point, directly against Hempel's argument. However, none of them could avoid the problem with Copernicus's heliocentric hypothesis. Of course, the test would have to be done based on the knowledge at the time of Copernicus. At that time, even if no one thought about dropping a ball from the Tower of Pisa, one could drop a ball from a building, a cliff, or a treetop. The test would show the ball hitting the ground exactly beneath the dropping point. Therefore, the Earth is not moving as predicted with strict deductive inference. The conclusion would be that Copernicus's hypothesis is false.

Scientists, most of the time, have to make judgments based on incomplete and uncertain information just as scientists during Copernicus's time about the heliocentric versus geocentric debate. Therefore, the recipe or algorithm offered by the H-D method would fail in deriving scientific

---

[2]The conclusion that the Earth is not moving and the Sun moves around the Earth is an obviously wrong conclusion even though it is based on the H-D model. Galileo conducted many experiments inside a ship. He found that although the ship was moving, one could not feel the motion inside the ship, the same as the ball dropped from the Tower of Pisa. Therefore, he invented the Galilean relativity.

proof. It does not work for scientists, period. Now, you may better appreciate the comment of Feynman that "philosophy of science is as useful to a scientist as ornithology is to a bird". But the conclusion should have been that the theories developed by the philo-scientia philosophers were not useful and not that philosophy of science is not useful.

The root problem with the H-D method is in its logic. The objective of the method is to *deductively prove the hypothesis*. The hypothesis in science should be a more general idea and not an idea to describe a specific case. Deductive reasoning can only go from general (idea) to specific (case) and cannot be used backward to prove a general idea. To prove a general idea based on specific cases, one has to use inductive reasoning as discussed in Section 7.2. Therefore, the H-D method is fundamentally flawed.

The H-D method may be the single most misleading concept developed by philosophy of science. It is harmful to those who are interested in science. Some people present the method as was originally proposed by Bacon or Descartes. As I discussed in Section 7.1, neither actually did. The origin of the idea may be earlier but is commonly considered best illustrated by Hempel in 1966 as I described at the beginning of this section. Therefore, the idea was not developed by scientists but a pure product by philosophers of science. With such flaws, especially after Popper's falsification theory, it is difficult for me to understand how this flawed method can have wide followers so that it permeates into the pre-college education and science management systems in the United States. Some people argue the positive aspects of the method by introducing the concepts of science to high school students. However, many students have taken these flawed ideas into their careers. One consequence is that now scientists have more difficulty with the science managers who learned about the H-D method in high school. These managers evaluate a scientific idea according to the hypothesis and not the problem the idea addresses. The damage has been produced to science and may take several generations to repair.

7.5 Causality

In philosophy of science, a large topic is devoted to causality. Theories of causation are supposed to explain the cause of an observation.

Philosophically, the law of causation says that things in the world are connected and cannot take place in isolation. The problem arises because there is no scientific law to determine causal relations. Some empiricists, such as Hume, are strongly against causality. They think of causation as suspicious because we cannot experience it directly, as we learned in Chapter 4, e.g., in Russell's chicken-farmer and Pavlov's dog-bell examples. In an extreme case, Hume refused to accept that the drop of a glass vase to the ground is the cause of breaking it. They prefer the term *explanation* to *causation*.

So, what is causation? Is it some sort of hidden connection between things? Conventional theories of philosophy of science spend much time arguing about cases such as when one drowned and what was the cause of the death. Was it caused by not being able to breathe or by not sufficient amount of oxygen in the blood? Or, when a person was near death after being bitten by a poisonous snake but had a fatal car accident on the way to the hospital, what was the cause of death? Was it the snake bite or the car accident? Or, when a man who has taken his wife's birth control pills does not get pregnant, are the birth control pills the cause of the man's inability to get pregnant? These discussions sound more like lawyers debating over an insurance claim case and are not what is in science. I do not think these cases are relevant to science or to philosophy of science. In principle, the first case demonstrates that a "cause" depends on the level of specifics of the discussion; the second case shows that when there are multiple causes, the time sequence and magnitude of the causes are what matter: and the third case draws attention to the importance of the relevance between the cause and the conclusion, an issue we will further discuss in inductive reasoning in Section 9.3.

In philosophy of science, we first eliminate the possibility that something can be caused by God.

A famous example of causality discussed in philosophy of science is the so-called flagpole and shadow problem. One can determine the height of a flagpole using the measured length of its shadow. In other words, one may state that the flagpole height is determined (caused) by its shadow length or the shadow length determines (results in) the height of a flagpole. There must be a flaw in this statement because the height of the flagpole does not change while the length of the shadow changes with time, and therefore, the shadow length cannot be the cause of the height of the pole. This example is used to argue for the asymmetry of causal relations. This is a valid point although this is not a very intelligent example because it is a result of misinterpretation according to the linguistic form of the sentence. This example may demonstrate that many problems discussed in traditional theories of philosophy of science have little relevance to real reasoning, but it illustrates the difficulty sometimes scientists encounter concerning what is the cause or result and how we define causal relations in science.

In our discussion, causation is defined as the common sense of the cause–result relation and not the one discussed in conventional philosophy or philosophy of science. There are a few forms of reasoning people use to determine the causation.

Causal relations in science can be unambiguously determined according to deductive reasoning, for example, if a process is described by a law that determines time evolution, such as Newton's second law of motion. Forces are the unambiguous cause of an object moving or not moving. The acceleration or deceleration of an object is caused by the net forces applied to the object, e.g., the forces in the direction of motion cause the acceleration and forces in the opposite direction cause the deceleration. Similarly, the cause of the electric or magnetic field can be determined by Maxwell's equations.

However, most often, a problem may not be described with a time-dependent law. For example, the ideal gas law or Ohm's law is assumed to hold instantaneously. In these cases, external causes are needed for any change. For example, when one pumps gas into a balloon, the increase in the pressure causes the balloon to expand which results in a decrease in pressure. Eventually, the balloon reaches a new equilibrium with a different size when the pressure is balanced with the outside pressure. Similarly, on the flagpole shadow problem, the cause is neither the pole nor the shadow, but the Sun. The length of the shadow is caused by the position of the Sun with time, not by the height of the flagpole alone.

In general, the causal relation in science has to be determined based on knowledge-understanding and/or deductive reasoning. Otherwise, it is considered speculations. For long-range causal relations, such as Darwinian theories, knowledge-understanding is required at a few critical points as well as the long-range gradual evolution.

Causation cannot be determined by inductive reasoning alone. Simple time sequences of the events/phenomena cannot be used as causal relations because they may be coincidental or they have the same cause, e.g., thunder and lightning both of which are caused by the discharge between clouds or between cloud and ground. When observations are not connected as a time consequence, e.g., statistical correlations of two parameters, their causal relations cannot be determined and should not be assumed known.

There is another form of reasoning that can derive causation. It is best illustrated by the following famous proverb:

"For want of a nail, the shoe was lost.
For want of a shoe, the horse was lost.
For want of a horse, the rider was lost.
For want of a rider, the battle was lost.
For want of a battle, the kingdom was lost,
And all for the want of a horseshoe nail."

The causal relation derived from this line of reasoning is that the cause of the loss of the kingdom is the loss of a horseshoe nail. The reasoning appears very regular and strict; each step appears to be possible, but the conclusion appears suspicious and problematic because the loss of a kingdom can be due to many reasons. When we examine the line of reasoning, we find each step is a smaller effect to a bigger consequence. For example, for losing a horseshoe nail, there is a range of possible consequences; but the worst was picked. In a sense, the logic is always from small to large. As we discussed in Section 7.2, deductive reasoning should go from a more general idea to more specific cases. Therefore, the logic is not strictly deductive. The reasoning becomes clearer when we realize that the loss of the kingdom is the result that is known, and that loss of nail is a speculation of the cause. The proverb describes a series of possible speculations for the loss of a kingdom. This is abductive reasoning! We know that abductive reasoning may derive flawed causation because one may be able to name many more substantial causes to explain the loss of the kingdom. Of course, one may defend the reasoning by the argument that there is no "if" in history. One then has to prove that every critical step mentioned in the proverb actually happened, e.g., showing that the chief general was about to win the fate-deciding duel but suddenly the horse fell because of losing a horseshoe with additional evidence of missing a nail on the horseshoe of the leg that caused the fall. In this case, the logic is purely observational and not deductive. We recall that observation does need proof by logic.

A similar example in science is the so-called butterfly effect according to which the flap of a butterfly's wings can cause an earthquake or volcano eruption thousands of miles away. This causality is based on an "analogy" to the chaos theory in mathematics in which a small difference in the initial condition can fundamentally change the mathematical solution of a system so that the solution becomes unpredictable. Chaos is an interesting mathematical theory to describe some complex nonlinear system or some critical processes at the critical point. However, given the mathematical theory, any small difference, not only that caused by a butterfly, in the initial condition is able to cause a dramatic difference in the solution. The conclusion should be that there are an uncountable number of potential causes, i.e., the system becomes unpredictable according to common sense. But why is the flap of butterfly's wing movement singled out as the only cause of an earthquake or volcano eruption? If it is truth, by which butterfly?

There was an earthquake followed by a volcanic eruption five days later in two neighboring islands in Indonesia in 2018; but one "geophysicist" argued for dissociation of the two events based on the butterfly effect, i.e., the earthquake could not have caused the volcano eruption but a butterfly wing flapping thousands of miles away did. If one traces a different effect than a butterfly, can't they conclude that a different effect causes the volcano eruption? Science is not history; one of its tasks is to derive a convincing cause of an observation. The butterfly effect, as discussed above, is based on abductive reasoning and is seriously flawed from a viewpoint of philosophy of science. It is based on a bad analogy to explain chaos theory. In science, holding a single fact and exaggerating its importance unlimitedly is a dangerous habit one should avoid.

7.6 Explanatory Inference

Often when we observe a phenomenon, we have tendencies to explain what caused it, e.g., what caused the extinction of dinosaurs. In this case, we know the result and (backward) predict the cause based on causality or analogy or personal experience/story. We find many possible potential causes just as there are many potential equations that have a known solution as you find that your friends have different explanations about the same observation/experience. An explanatory inference assumes that a specific cause for the known result is based on abductive reasoning. We know from Section 7.2 that abductive reasoning is deeply flawed and is not scientific reasoning. To eliminate competing possibilities, the theory introduced the concept of inference to the best explanation or IBE. But what is the fundamental reasoning employed in IBE? Philosophers have different opinions. Some philosophers describe IBE as a type of inductive inference because it is not deductive reasoning. Others think it is not inductive because the conclusion is specific and not a generalization. As we see, either misses the mark because IBE is abductive reasoning. However, this IBE produces a new problem: How can one determine the *best* potential cause, since we know that one's best may be the worst to others?

One of the models that specify the IBE is the "covering law model", aka the "deductive-nomological (D-N) model", in philosophy of science. According to it, what constitutes an explanation is that (1) the connection between the premises employed in the explanation and the conclusion has to be deductive, (2) all premises employed have to be true, and (3) the premises have to include one general law. Let us learn how it works. For example, one may have observed the fact that the interest rate is almost always positive. Some economists proposed a theory based on a "law" that people prefer immediate and certain consumption to future and uncertain one. According to the "law", the one who wants immediate consumption should pay those who defer the consumption. Therefore, the interest rate is always positive. This does sound like a reasonable explanation of the observation and many philosophers of science agree (Alex Rosenberg, 2005). However, the "law" in this case seems inconsistent with the law of demand and supply in economics and hence the "law" itself may be questionable. It is simply a rationalized guess and may not qualify as a law. Furthermore, because it is invented to explain the observation and does not have wide applications, the reasoning appears circular. An alternative explanation would be associated with the risk faced by the lender.

The borrower may not be able to pay back or the lender may need to make a substantial effort to get the money back, e.g., by hiring a debt collector. A positive interest rate is needed to provide the incentive and to mitigate the risk to the lender. The interest rate can be considered as a charge for the risk. Obviously, most people would agree that the second explanation is more reasonable although no law is involved. Therefore, the covering law model does not seem to work.

A more commonly accepted way to determine an IBE by philosophers is Occam's razor, i.e., the *simple* or parsimonious explanation is IBE. Let us look at the example of the extinction of dinosaurs. One explanatory inference is that dinosaurs were killed by a powerful disease; this is very simple and very likely. However, Luis and his son Walter Alvarez in 1980 proposed the Alvarez hypothesis that the extinction of dinosaurs was due to an asteroid impact 65 million years ago. The idea was based on the observation of an abnormal, but unquestionable, layer of sediment of chemical elements that are commonly seen from asteroid impacts in the Earth's crust. The age of the layer coincides with the period of the dinosaur extinction. Then, one can develop physical models to estimate the aftermath of the impact of a sizable asteroid. The dust cloud produced by the impact blocks the sunlight so that the world becomes colder. Given the known conditions for dinosaurs to live, warm and humid, from other studies about dinosaurs, the environmental conditions for dinosaurs to live could have been destroyed so that they all died. The reasoning for this theory seems possible but very complicated, especially compared with the disease theory. How can we determine which one is IBE in this case?

This example may be used to demonstrate how to analyze the logic in science. The Alvarez hypothesis was first categorized as inductive reasoning by some conventional theories of philosophy of science because it was formed from some known facts to generalize to an unproven conclusion. But it does not seem to be inductive logic because the conclusion is very specific (an asteroid killed the dinosaurs) and not merely a generalization because there is no other case to generalize from. On the other hand, classical deductive logic is from a (general) cause to predict the result(s) and the prediction is to the future and not the past. The reasoning for the Alvarez theory is from a known result (extinction of dinosaurs) to infer the cause so that the *prediction* is made about the *past*. Therefore, logic cannot be deductive reasoning either. We now know it is abductive from the discussion in Section 7.2.

Most scientists in the 1980s would talk about the Alvarez hypothesis as an interesting idea among a few other ideas. It wasn't until thirty years later, in 2010, that the hypothesis was endorsed by the science community and became knowledge. If you were a scientist studying the subject during the thirty years, you would have to make your judgment, based on the few pieces of information, on whether you would support or be against the idea. As a scientist, your correct action should try to find relevant information regardless of whether to support or be against an idea.

After the discovery of the Chicxulub Crater in the Yucatan Peninsula, which is now considered the impact site, the impact time was determined in 2013 to be 66 million years ago. Ordinary people may consider 66 and 65 million years as close enough. They may second guess the reason for the 30 years of delay. In fact, during the 30 years, scientists have provided evidence for the asteroid's impact, its global effects, and, eventually, the impact site. With these pieces of evidence, the Alvarez hypothesis from one of the inferences became IBE and eventually knowledge-understanding while the other ideas gradually died out. It is no longer a pure hypothesis. The proof is also no longer based on abductive reasoning but based on the combination of scientific inductive and deductive reasoning, starting with the impact and then effect and consequence of each step after it. Therefore, it was scientifically correct not to consider the Alvarez hypothesis as knowledge when it was first proposed, and it is correct to consider the theory knowledge after more evidence was provided for each key scientific idea. This case may explain the difference between the science process and the knowledge it eventually produces. Also, note that the final decision of IBE and knowledge in this example is not according to simplicity. We will further discuss this problem in Chapters 8 and 9 because many people, including many philosophers of science, confuse science with knowledge.

I listened to a class by a philosopher teaching IBE. To the professor, an IBE is obviously correct because experts have told us so. However, if you were an expert, you would be the one people are asking for judgment. In this situation, you may find that an explanation has to be based on knowledge-understanding and not merely knowledge-that or science.

But you may not have a firm answer because the information is insufficient. For example, that zebras have stripes is knowledge-that. It cannot be directly used for explanation or understanding why zebras have stripes. There have been explanations, such as "wearable air conditioner", camouflage to confuse big predators, and the latest, camouflage to confuse flies. It is clear that every explanation has to involve at least another knowledge-that that is relevant and reasonable to the explanation. Are you able to provide an IBE as a firm answer, bearing your name? If you were a good scientist, you would present all three possibilities, not an IBE.

Philosophically, IBE does not define what should be used as the basis of an explanation and hence is a flawed method if used in science. People are different and believe different things to decide IBE. We have demonstrated that the IBE provided according to the simplest explanation (a powerful disease that wiped out dinosaurs) or covering law model (reason for positive interest rate) can be fallible. Therefore, the reasoning for IBE cannot provide a scientific conclusion or be used to prove a new scientific idea. Historically, there have been too many IBEs that are false as documented by Larry Laudan (1981). The goal of science is seeking truth and inventing knowledge. IBE is simply not trustworthy to scientists. Furthermore, scientists do not sit there every day to explain phenomena using a known IBE or covering law; even if it works, the explanation may not lead to new knowledge. To prove a new scientific theory to be true is much more difficult than to explain a phenomenon with an existing theory. In conclusion, IBE is not useful to scientists in scientific research although it may be useful to the general public.

7.7 Against Scientific Methods

Paul A. Feyerabend questioned the concept of the universality of the "scientific method" in his book *Against Method* in 1975. "Everything goes!" he declared. Later, physicist Lee Smolin wrote an essay, "There Is No Scientific Method", in 2013, and Daniel Thurs, a historian of science, published a book *Newton's Apple and Other Myths about Science* in 2015, both of which concluded that the scientific method is a myth or, at best, an idealization. According to Feyerabend's arguments, there is no scientific method, which would result in scientific anarchy. Interestingly enough, on the other hand, philosophers Robert Nola and Howard Sankey, in their book *Theories of Scientific Method* in 2007, argued that Feyerabend, despite the title of *Against Method*, accepted certain rules of the methods. They attempted to justify those rules with a meta methodology.

Feyerabend opposed any system of rules and constraints in science because science is an aspect of human creativity. Creativity should be encouraged and not be constrained. He thought that great scientists are opportunistic and creative. He was strongly against Kuhn's promotion of normal science and thought that the concept of the paradigm never succeeded in controlling science. He thought that specialization turns scientists into "human ants" being unable to think beyond their training. He used the case of geocentrism versus heliocentrism from the history of science to demonstrate that the most popular scientific method fails, a nightmare that has haunted philosophers of science ever since. There is no need for discussion on this case — if we had followed a rule, we would have considered Copernicus wrong, he concluded.

Without rules, however, how can science proceed? Or, does science need to be rational? Feyerabend provided two principles to follow: the principle of tenacity and the principle of proliferation. The principle of tenacity is that a scientist should not give up too easily. Most good scientists would agree with this first principle. It is common that even if an idea has been rejected, the scientist who invented the idea would not forget the "creative idea", because they have put much effort into the idea and, maybe, the theoretical or mathematical development of it. When learning about some new and relevant results, they would tend to make connections to see whether the rejected idea could be revived. Or, when one starts working on some new problem(s), they would naturally think about whether the old ideas or mathematics could be useful. Therefore, the first principle is quite natural to scientists. There is nothing so special about it. However, there is a problem whether it should be counted as a principle.

For the principle of proliferation, he envisioned that science should be like a marketplace of new ideas, either privately or publicly. This could be problematic to most scientists because, as I explained, scientists have too many, not too few, "creative ideas" themselves. The problem is how to eliminate and reject most of them. If all ideas from everyone were published, a science journal would not be different from a junkyard. Feyerabend clearly caught the wrong problem. Most conventional theories of philosophy of science do not appreciate the importance of a scientist's credibility, an issue to be discussed in Section 10.3. It is true that science has a function like natural selection. However, this is to select only from the best products, but never from everything by every contributor. People who keep proposing "creative" ideas without being able to prove them will eventually be filtered out of the science community, i.e., losing their jobs as a scientist.

Then, what governs science activities or processes? This is a more interesting question. Given that the science community is diverse with many completely different or opposite beliefs, views, and ideas, if there is no commonly agreed method, what do people agree upon? Is it really scientific anarchy as Feyerabend described? Just imagine this situation: You think that you are very intelligent and knowledgeable, you have worked very hard for three years on a science problem, and you developed an idea that you think would produce an earthquake in your science community. You then submit a paper to a science journal for publication, but it is rejected. Wouldn't you be extremely upset and frustrated? But after reading the referee's reports, you would agree with the referees! What is going on here? How could you agree with someone who says your effort of three years was all wrong, without a commonly agreed upon scientific method? Amazingly, such a chaotic science community consisting of people who are intelligent, rational, and independent, but often with opposite ideas, successfully and continuously produces uncountable new ideas and knowledge. What makes this happen? It seems that all scientists agree upon a certain thing. What is it? If I had learned about it early in my science career, my scientific endeavors would have been much more successful and less stressful.

Now, as we can see, science is more anarchist than democratic. No one has absolute authority, but a majority vote is not taken as a justification of correctness. In fact, leading scientists are often leaders with opposite scientific views. Scientists are not allowed to speak nonsense; all points of view are scrutinized and debated upon constantly before and after the publication of a work. But how do scientists know who is right or who is wrong? Then, science should be in chaos, should it not? What makes science different from other fields is what we call scientific reasoning. Scientific reasoning is the most important aspect of philosophy of science. Although deductive logic and inductive logic are introduced in every book of philosophy of science, the concept of scientific reasoning is rarely discussed. I will be introducing and discussing this in Chapter 9 to provide a guide to this learning process and answer the question of what all scientists agree upon. But for now, what really is science?

Questions for Thinking

1. Have you learned scientific methods before? If yes, what method do you think is most useful or powerful? Please describe it.
2. Which methods discussed in this chapter have you used before for predictions? Which methods do you think are most useful? Have you thought about the potential flaws of each method?
3. Please name the reasoning that you think you have used. Which reasoning can bring truth and which may bring flaws, in your opinion?

Chapter 8

What Is Science?

8.1 Definition of Science

With so much discussion about science, we have not defined it yet. As we learned in Chapter 1, "science" is from the Latin word *scientia*, meaning for results of logical demonstration that reveals a general and necessary truth. As a counterexample, we have learned about Einstein's relativity being based on an assumption that was illogical at the time according to Galilean relativity. So, was Einstein's relativity not science according to the original definition of science?

According to Encyclopedia Britannica, science is "any system of knowledge that is concerned with the physical world and its phenomena and that entails unbiased observations and systematic experimentation. In general, a science involves a pursuit of knowledge covering general truths or the operations of fundamental laws." According to this definition, science is both the *system of knowledge* itself and the *pursuit* of it. According to this definition, Einstein's relativity was a pursuit of knowledge and should be classified as science. However, in this definition, science is only for the *physical world*, i.e., not the social or psychological world. Therefore, research fields other than natural sciences cannot be science. Furthermore, it is somewhat unusual that in this definition, the keyword "logical" is absent. Does this mean that logic is no longer a requirement in modern science? Is science rational if logic is not required? As we have discussed, knowledge-understanding is often NOT based directly on observation but on a specific interpretation of it. Interpreting observation includes a rationalization process. Also interesting in the definition is that knowledge and truths are considered two distinct concepts, consistent with the general concepts developed in this book.

To the public, the reason why science can reveal the truth is that science is rational and proves with evidence. However, to a scientist, it is not clear whether rationality alone can definitely reveal the truth. When two scientists debate intensely and disagree on the interpretations of an observation, can they both be rational? Each interpretation is based on its own model or idea. Both appear rational and provide "logical demonstrations" but differ by their logic and rationalization because of incomplete information. Which rationalization reflects the truth? Who is unbiased? Is the interpretation we choose "unbiased"? How do we know? Therefore, any definition of science with "unbiased", "rational", or "logical" can be questionable to scientists because these terms require complete information. When information is partial, one thinks an idea is rational and logical, but others could think it irrational and illogical. Science has to be rational (based on reasoning) in order to differentiate from religion but rationality itself does not guarantee to derive knowledge that reflects truth when information is not sufficient and/or when there are multiple choices of rationalizations. Science is defined by some as a general understanding of how humans gain knowledge (of the world) and an understanding of what makes the work that descended from a scientific revolution different from other kinds of innovations (Godfrey-Smith). This definition takes science as a set of understanding, but science is much more than understanding as we discussed above.

With consideration of the above issues, we are ready to define science. Science is a process on the whole societal level aimed at inventing new knowledge to systematically build, organize, and validate true beliefs based on observation, conceptualization, and reasoning in the forms of evaluable explanations and predictions while filtering out the knowledge that has been falsified based on observations. In a simple sentence, science is *a process of developing knowledge*.

Let me unpack this complicated definition as philosophers usually do. Science is a *process* to develop knowledge. Although sometimes the word may refer to the enterprise or new results of science in general, it is mostly about the *process*. "We are *doing* science", for example. It is important to note that most of the scientific results will not become knowledge or they become knowledge in very different forms from their forms in science because there is a substantial instillation, crystallization, reduction, condensation, and abstraction process from science to knowledge.

Science searches for underlying truth that connects observations (knowledge-that's) and conceptualizes the connections with *reasoning* based on existing and develops new knowledge-understanding. Observations here include systematic and/or controlled experiments and experiences that can be quantified. Since knowledge could be wrong, results during the science process have more chances to be wrong than knowledge does. It is fair to say that a significant fraction of the results in science is either incorrect or incomplete. The review and referee processes involved in scientific publications and presentations of a scientific result may be able to substantially reduce the chance of it being false, which will be discussed in detail in Section 8.3. Therefore, one of the biggest misunderstandings by the public, misled by the conventional theories of philosophy of science, is that science is the same as knowledge or truth. Far from it. An essential function of science is to gradually filter out false ideas. It is equally important to know that previously falsified ideas could possibly become knowledge when our knowledge and information improve. In the process, evaluation provides quality control of the candidates of ideas.

The science process is open to the whole of society. This is fundamentally different from other processes or ways of gaining knowledge. A "scientific idea" has to be subject to open debates, scrutiny, and criticisms by the whole of society; otherwise, the idea is not scientific. For example, after the result developed in isolation is open, say, found from a coffin or reported from carefully held secret, it takes time to falsify, justify, or modify it in accordance with the scientific knowledge at the time in the scientific process. Therefore, science has a strong social attribute. The curiosity and beliefs held by individuals/small groups are the primary drivers of science. Results of individual curiosity in isolation do not account scientific results until they are open to the public and go through a falsification process by the science community and anyone who is interested. On the other hand, scientists, especially good ones, are individual thinkers. They are not influenced by their peers although they will listen to and considered others' ideas carefully. Each scientist makes scientific judgments based on scientific reasoning and observational facts as will be discussed in Chapter 9, and not on a majority vote.

The science process includes three major pillars or elements: observation, conceptualization, and evaluation, as discussed in Section 5.1. Without these three elements, it is not science. Science aims at inventing new knowledge and at the same time filtering out the false knowledge, and aims less at applications of existing knowledge except if the application is to invent more (detailed) new knowledge. Pure application is more important for engineering and other applied fields. The boundary between applied science and applied fields may be fuzzy. For example, building a new bridge may involve some new aspects, although the fundamental theory governing the stability and durability of the bridge is knowledge and most "new" designs do not involve much new science except for a completely new concept.

Science is not defined according to rationality or evidence or method, as defined by the traditional philosophy of science. Rationality is a key feature in many definitions to distinguish science from religion. However, we have learned that rationalization involves personal judgments when information is partial; each scientist puts a weight function on the available information. Therefore, rationality alone does not necessarily lead to knowledge that reflects the truth in science. What all scientists agree upon is *scientific reasoning*. Our definition tries to avoid using the term "logic" because it refers to specific *forms* according to logicians in philosophy of science who have dominated the field for over the last hundred years and produced many misleading ideas and theories. Also, the definition avoids using the word "evidence" because it often depends on the interpretation of a fact based on a personal rationalization. A single observation can be interpreted differently to serve as evidence, very often for opposite ideas. Therefore, the definition of science removes the ambiguity caused by different interpretations. The debates over these different interpretations, however, form the core part of the scientific process.

The ultimate judge of a scientific idea is observation/experiment/ experience. However, existing knowledge, conceptualization, evaluation, and reasoning also play significant roles in judgment. In this sense, extreme forms of rationalism are incompatible with science. Existing knowledge is based on previous experience although it may not always correctly reflect the truth. However, when inventing new knowledge in a science process, existing knowledge on the subject must be carefully justified, modified, or falsified. Extreme forms of empiricism would be a handicap in science as they do not recognize the importance of conceptualization and reasoning. Newton's theory, Einstein's theory, and Maxwell's equations cannot be derived solely directly from experience. Because observation and experience may depend on interpretations, there is a large uncertainty that limits falsification as well as justification of an idea in science.

In science, many ideas are proposed and evidence for each idea is searched for, presented, tested, and evaluated. Most ideas/models during science will be filtered out or substantially modified. Those filtered-out ideas are not considered knowledge although they are often reported in the news or make big headlines.

The important processes from science to knowledge are reductions and abstraction. Knowledge is required to be relatively simple so that it can be used as the basis for the public in decision-making. It has to be connected, or "systematic", very much like clusters of grapes, so that people can easily remember, master, and apply it quickly to a wide range of issues in everyday life. It "makes sense", but "sense" — or intuition as it is often referred to by scientists — itself can evolve according to new knowledge. The internet has become a source of information for everyday life. It may be a convenient place to dissimilate or search for existing knowledge, but we have to be very careful with what we learn from the internet since not all information is knowledge.

To distinguish science from non-science, we use *evaluability* that requires any comparison, prediction, or interpretation to include some quantitative flavor. We do not use "quantification" because some fields of research may have not reached the level of rigorous quantification. However, any theory or observation has to provide some assessment on orders of magnitude so that it is not pure speculation. Evaluability is a stricter requirement than testability which is often used in theories of philosophy of science. It eliminates the possibility of the supernatural, religious, and other irrelevant ideas, although an irrelevant idea may later become relevant when more information is available. This definition is different from Popper's concept of demarcation. According to our definition, science includes processes that *move toward being testable and then evaluable* so that it potentially includes many fields of research, at least part of them, such as psychology, economics, and sociology, which are defined as pseudoscience according to Popper's demarcation. However, these fields need to include observations, conceptualizations, and evaluations.

One may argue that because Darwin's evolution theory did not involve any mathematics or evaluation, evaluation may not be an essential feature of science. However, this argument is flawed. Darwin's theory started as a speculation, and it became a scientific idea when scientists made measurements based on the dating method and put different pieces of evidence reasonably on the timeline and family tree of evolution. Without evaluation, an idea or a field may be referred to as speculation and is not science. According to this definition, although science does not require that predictions be completely consistent with observations, some level of quantitative control is required. The scientific process allows modification of an idea when there is a difference between the prediction and observation.

Evaluation plays a key role in modern science. Both observations and theories have to be evaluated quantitatively to a certain degree so that most pure speculations can be eliminated from the possible candidates while systematic quantitative observations can provide hints or clues toward the invention of a theory. Without evaluation or quantitative control, knowledge-understanding cannot be separated from pure speculation. Speculation with observation and scientific reasoning can be a starting point of conceptualization. It may gradually become science and eventually knowledge when evaluation is included, such as in the examples of Darwin's theory and the Alvarez hypothesis. Some research efforts provide massive observations but with little conceptualization, such as the observations of behaviors in elevators discussed in Section 4.5. They cannot be classified as science until better conceptualizations are developed. Dream interpretations cannot be science at this time because there is no quantitative control. This reason is in contrast with Popper's demarcation discussed in Section 4.5. The inclusion of evaluation in the definition of science distinguishes science from other types of guesswork and investigations. Mathematics and numbers are an essential part of science;

however, including numbers alone does not necessarily qualify an idea as scientific because science needs to bring new understanding.

Science, as a way to develop our knowledge, was originally defined for natural subjects. Its success in rapidly increasing human understanding of the natural world and problem-solving capabilities has made science a model to emulate by other branches or disciplines of inquiries so that many academic branches or disciplines name themselves as "science". We welcome this trend; however, it has caused confusion. For example, some define science as "Evidence-based" studies. A potential flaw and confusion in this definition concerns how to generalize a theory that is based on a singular case or ill-constructed statistic studies and what can be counted as evidence. Referring to science as evidence-based studies was originally used to distinguish from religious or superstitious approaches. However, now there is at least one group that named itself "science of religion". We have learned that a fact itself can be very different from the interpretation of it. Almost any theory can find some examples and/or phenomena that can be used or interpreted as evidence for it. In other words, "supportive evidence" *alone* cannot be used to prove any scientific theory. This problem has been recognized by Popper, although I think he went to the extreme on this great idea so that it became flawed.

Is there a non-science process to gain knowledge? Can human beings gain knowledge, not through a science process? Yes, if the process is not based on observation, conceptualization, and reasoning or the idea does not go through scientific debates and criticisms and is not controlled quantitatively. For example, in some studies, knowledge may be produced by Gods or according to the beliefs of leaders or a powerful group of people. Similarly, if van Leeuwenhoek's observations of microbial lives or Mendel's plant hybridization experiment did not go public, their results would not be scientific. Many fields in art and literature do not involve evaluable explanations and/or predictions and, therefore, they are not science. Philosophy and philosophy of science are fields with open debates and invoke observation and conceptualization. Lack of evaluation makes them less able to filter out the false knowledge and keeps the false knowledge for some reasons. Some "knowledge" is not true knowledge-understanding (although it can be knowledge-that); it may be believed by many people. In some fields, individuals may falsely take their ideas and beliefs as knowledge without open debates, especially before the modern era when large-scale communication was difficult.

8.2 How Is Science Organized?

Now, we examine how science is organized and operates. According to the discussion of scientific paradigms, science has a more hierarchical structure from more general branches of science, e.g., physics, mathematics, chemistry, and biology, to more specialized ones which we refer to as disciplines and subdisciplines, as discussed in Section 5.5. Scientific paradigms at higher levels provide organic connections among different lower-level paradigms. There are also interdisciplinary research programs. Scientific research takes place most actively on the subdisciplinary or sub-subdisciplinary level which is highly specialized. At the fringe of a paradigm, science may either further expand or may have reached a known limit.

The area beyond a paradigm needs to be covered by something else which could be either higher-level paradigms that cover multiple disciplines or a new paradigm to be developed. For example, Einstein's relativity and quantum mechanics cover Newtonian mechanics for situations when the speed of an object is close to the speed of light or if its mass is highly concentrated and for objects that are extremely small. Here, we should emphasize that Newton's theory is applicable under the specified conditions and has not failed in the domain it covers. Although relativity and quantum theory have so far been successful in science, it would be difficult to predict what the theories corresponding to relativity and quantum theory would look like two thousand years from now because of new observations and new understanding invented during the years. It is not impossible that the two theories are unified then. If this does happen, it is a clear case that the present forms of both theories are falsified or partially falsified. However, even if this happens, Newton's theory would most likely hold in a form similar to the current one under similar conditions that are specified. I do expect Newton's theory to survive history in a manner similar to the plane geometry doing today although the latter is more than two thousand years old. This is because Newton's theory has been tested quantitatively extensively under the specified conditions similar to plane geometry two thousand years ago. The statement "there is no such thing as Newtonian gravity which acts at distance; rather space is curved" (Rosenberg, 2005) may not be intuitive enough for the general public to determine the gravity that they experience in everyday life in non-relativistic situations. For example, one has to explain intuitively and quantitatively how to determine the curved space when dropping a ball from the Tower of Pisa. I expect relativity, not Newton's gravity, to be modified in order to become knowledge-understanding regarding gravity.

Each branch should have an independent department in a mid-sized university or college. Depending on their maturity, each discipline may have standard college textbooks as the requirement to study, which are the material form of knowledge, as explained in Chapter 3. Commonly used textbooks are relatively rare at the subdisciplinary and sub-subdisciplinary levels since ideas in the field are still fluid. The development of a standard textbook may be an important milestone for the field reaching maturity.[1] Most research results, or knowledge in the science phase, are published in scientific journals. These journals also periodically invite leading scientists to write topical reviews for a broader audience. These topical reviews may be the most useful source of information for newcomers on the topic. However, standard textbooks disconnect students from the original journal publications of many classical works. As described in Section 3.1, the results from the original publication have been abstracted, condensed, and reduced to *knowledge* in textbooks. They are more accessible to a greater

---

[1]For example, with the growth of the discipline and increasing demand for education and training, the first textbook on Space Physics was written by a group of leading scientists; each wrote a chapter on the topic in which they specialize. This textbook provides good coverage on various topics. However, it isn't easy to use for teaching because the transition between chapters is less smooth. Several books were also published; each emphasizes a few aspects of the discipline with a sketchy description/coverage of the overall picture. Some contain flawed statements and concepts in areas in which the author has less expertise. The first textbook was rewritten to become a more coherent one more than two decades later. Since these textbooks are a transition from science to knowledge in a discipline that is still under development, an interesting issue for the authors is how to decide what scientific understanding has reached the level of knowledge.

audience and at the same time are critically reviewed by each instructor and student of the course.

In a mid-sized university, within a department, i.e., a science branch, there are only a few faculty members in each discipline and much fewer in a subdiscipline. This is often an insufficient intellectual pool of expertise (often referred to as a "critical size") to carry out an active scientific research program. Therefore, most scientific activities do not take place within each mid- or small-sized university. Scientists have formed various national and international societies, associations, or unions as the hub of scientific activities organizing regular conferences and workshops and sponsoring scientific publications.

There are conferences at each level. Conferences of each major branch of science can attract attendances over a few thousand or tens of thousands where one may be able to listen to lectures given by the icons of the field and learn a broad range of new ideas. However, smaller-sized meetings and conferences may provide better opportunities to learn the details of a particular topic of interest. Topical conferences and workshops, of a size of around 100 people, provide the most desirable environment to learn and exchange ideas.

There is more than one international science journal at and above each discipline level, where readership increases with the level of coverage. Publications in a high-level journal are less specialized while having a higher impact. However, for details of a new idea, one has to go to lower-level journals which provide a space for nut-and-bolt information if one is really interested in the idea. Unfortunately, because of the introduction of the "impact factor", a flawed index to measure the significance of a science journal, top-level journals have been using "newsworthiness", instead of scientific merit and novelty, as the primary criterion for selecting works for publication. This type of indexing is flawed because it is based on the number of citations of a paper acquired within a very short period. Newsworthiness is strongly influenced by the interest in society at a given time. For example, during the COVID-19 period, a discovery made from a new science satellite mission was judged by a leading science journal as unimportant. In contrast, a personal opinion, which contains little real substance, of a COVID-19 expert was judged as important by the same journal. With such a measure, there is no question that major newspapers would be ranked even higher. The problem is that few people will still remember the news stories a few years later. In my opinion, the impact of a scientific result should not be measured solely by the number of citations received in a *short* period of time. After all, a science journal is not a newspaper.

Kuhn's definition for a paradigm requires it to "attract an enduring group of adherents". He did not specify how many people are needed to qualify as a "group". A discipline may consist of thousands of active scientists worldwide that support and sustain a decent journal and regular annual conferences. A subdiscipline may consist of a thousand scientists. When the size of a subdiscipline exceeds this, it is more likely to split up into more subdisciplines. In such a science subdisciplinary community, one would not be surprised to name two or three different general approaches, each attracting well more than, say, 300 active members. These 300 scientists have a more or less general agreement on many common ideas to distinguish their ideas from or oppose a different approach. They, their students, and their students' students could work for a few generations on the same subject and so do their opponent groups, noting that scientists in other research institutions and national labs all received education in universities. If groups of this size are each qualified as a paradigm, a paradigm can change relatively more frequently and easily. Each of these paradigms is much smaller than the revolutionary ones, such as the Newtonian paradigm, as Kuhn originally described. Therefore, Kuhn's second condition for a paradigm, i.e., open ended with plenty of problems for the group to resolve, may be more important for the definition of the paradigm. In practice, each paradigm continues to hold while the group drifts from one topic to another when some new scientists join it and some leave, mostly determined by the topic. Some people characterize such a paradigm as a tribe. Indeed, on their appearance, believers of a paradigm may behave like tribal members. For example, they may discuss the strength and weakness of their paradigm as well as the opposite one and identify the best evidence for each argument. However, the fundamental difference between a tribe and a science group is that tribal members are related by livelihood but the paradigm is based on common scientific beliefs and interests. Scientists adopt a paradigm voluntarily.

If science and engineering are distinguished by new knowledge invention and practical applications, the science phase may last during the period when active science development takes place until it is handed over to engineering. But such a narrowly defined science field would challenge Kuhn's statement that only a single paradigm dominates a field. One may argue that these approaches at a subdisciplinary level are not paradigms. However, each does satisfy the definition of a paradigm and they do exist in the science community in a manner similar to the USA and USSR during the Cold War as discussed in Section 6.1. Each group organizes sessions in a conference targeting their constituencies and some refuse to referee manuscripts from the opposite paradigm. This is not a negative scene in science because it does avoid many unproductive debates which cannot reach agreements on anything, mostly because of incomplete information. This stable situation can only be broken by new results that conclusively resolve the root differences between opposite paradigms.

The modern science system is an honors system. Bad behaviors are condemned, and bad players often have to leave the profession.

8.3 How Do Scientists Communicate?

To the public, the issue of knowledge is about whether an individual knows or does not know. In contrast, to a scientist, the issue is whether an idea is correct/trustworthy or not. Communication among scientists is about the details of proof or rejection of an idea. It is true, as described by Kuhn, that most scientific research nowadays is published in professional journals which are difficult for non-experts to read. Even an experienced scientist would have tremendous difficulty reading a research article from a neighboring field. A non-expert would find it very difficult to appreciate the progress of a field. In an extreme case, some non-experts even claim that the scientific results are made up and not real, as will be discussed in Section 10.3. This is not due to language problems, but rather background knowledge. To fully understand a scientific term, a non-expert may need to listen to a full lecture or read a relevant chapter of a book. It is clear that there is no way to require a scientific paper to be easily understood by non-experts while maintaining its scientific rigor in a limited length. If the requirement for science rigor loosens slightly, the scientific contents can be appreciated much more easily. This type of presentation can be found in science seminars and tutorials of a topical workshop. It is understandable that a philosopher, historian, or sociologist would consider the scientific progress in a science field "minuscule" (Kuhn, 1962), but this is how human knowledge is accumulated before condensation and reduction to knowledge-understanding. The difference is due to the time scale of measurement. The active life of a scientist may be less than 50 years in a science field, but philosophers and historians measure the science progress in hundreds or thousands of years.

In scientific publications, there is a system designed to handle and control the review and recommendation process, a falsification process. In a peer-reviewed journal (NAS, 1995), a system proposed by philosopher Henry Oldenburg (1619–1677), as opposed to free internet-based journals, a submission is refereed by leading experts on the subject. The submission and referees' recommendations are judged by a board of editors who are chosen by the science community according to their scientific qualifications, scientific integrity, and fairness. In this referee process, the author and referees can have multiple rounds of comments and rebuttals. Note that the process is under the watch of the editor, so both sides have to behave professionally as scientists. Abnormal behaviors, on either side, can be spotted by the editor. Repeated abnormal behaviors will eventually damage one's scientific reputation and can negatively impact the career. Authors have the option to request the exclusion of certain scientists as referees, but a sound scientific result should be able to defend itself no matter who the referees are. Exclusion of someone as a referee is the last thing an author should request because it is not an image for a sound scientific result.

In most credible scientific journals, a manuscript is reviewed by two or more referees. A knowledgeable and impartial referee will provide a candid assessment of the manuscript to the journal editor. Very often, authors can learn a great deal in such a referee process. Sometimes, the referees can be more straightforward or, in other words, less polite. For young scientists, this could be hard to take. Sometimes, after receiving a referee's reports of this nature, a younger scientist could feel personally insulted. A few young scientists have complained to me about this. They make guesses about who the referees were and what their motives may be, although often the guesses are wrong. However, this is normal as long as the referee's reports provide scientifically justified arguments. The author should not pay attention to how the comments are worded. What the author needs to do is to reexamine their work and strengthen every point that has been mentioned. In practice, the editors of science journals can play a crucial role in dampening destructive behaviors by selecting referees who are able to make sound scientific judgments while being relatively moderate and neutral for a controversial manuscript.

Quite often, the comments could result from the unclear presentation of the author's idea(s). The author has to read the manuscript from the reader's point of view. One has to remember that the author knows exactly what is being presented, but the reader does not necessarily know. A small jump in the reasoning could lead to great hardship on the referee's side. A young scientist may underestimate the difficulty when a referee comments on an unclear presentation. Usually, unless the referee is absolutely sure about a flaw in the presentation, they do not want to make false accusations that may damage their own reputation. If an expert referee could not understand exactly the reasoning presented, how could an average reader understand it?

How can scientists from different parts of the world, with different cultural backgrounds and expertise, communicate peacefully at science conferences and during publication review processes? Presented in conferences and publications are observational results, interpretations, theoretical models, analytical and numerical details, or consequences of theoretical models, as well as a description of scientific instrumentation/ experiment and their validation. What is included in these diverse presentations and publications is evidence or proof for the authors' belief. There is no well-defined scientific method or format, so to speak, for all presentations. Also, its final version in the publication is often not the thinking and/or logic the authors originally held toward a particular problem. For example, frequently, the author starts an idea with abductive guesses, but the final proof has to be presented with evidence connected to deductive and inductive reasoning. It includes what the authors learned from various successful and unsuccessful tests and investigations with modifications in order to withstand the criticisms or falsifications from the audience of the conferences and the referees of the publication.

In the referee process, not only do the authors change or modify their views but so can the referees if the authors can make a strong enough case for their ideas. Because the review process is oversight by the science community and by the editors, quite often, a referee is convinced and accepts a work that is negative toward their own ideas. Is it amazing if one assumes that the referee holds more power? This is how science works! On the other hand, an author may also be convinced that the criticism from the referees is valid and reasonable so that the manuscript is withdrawn from or rejected by the journal when warranted. Given that both sides of a submission, i.e., the authors and referees, are experts on the subject and are able to put their personal biases aside, the outcome from the peer-review process with the judgment of the editor would be accepted relatively peacefully by the two sides in general. The fact that a scientific community can withstand tensions such as this and can continue prosperously may present a key to understanding what science is. Again, what happens during these exchanges between the presenters and audience as well as between the authors and referees? What is a consensus based on?

What makes the scientific review process work is that scientists have to, at least, speak the same "language", so that they can understand each other and agree on something in common. If they speak different languages, they will have difficulties communicating or even debating. The language here clearly is not in the linguistic sense and does not refer to the words or terminologies they use, but rather the reasoning used in the exchanges and debates. Scientists learned reasoning consciously or subconsciously through their training. They accept a few forms of reasoning among the forms discussed in Section 7.2. I call them "scientific reasoning" and will discuss this further in the next chapter and reject other forms. The author of a manuscript has to present the work in a clear line of reasoning which may be different from the line of thought when one derives the results. The task of the referees is to identify any flaws in the work and in the line of reasoning based on the current understanding of the problem and to help improve the readability of a manuscript.

Can we trust a referee and the falsification? When the personalities of the author and referee are involved, sometimes, it is difficult to separate science from personal biases. An idea could be criticized because the referee does not like the author or the referee may be defending older theories and refuses to accept new ideas. Science relies on the self-regulation of the honors system. Most scientists have learned the code of conduct in their training and early career. However, sometimes, anomalies do occur in the review process. The good news is that a single decision does not determine the fate of a study because there are multiple journals to submit to, which compete for the best and most important results. It is difficult for a top-ranked research result to be rejected by all potential journals. In rare cases, the authors may bear some blame, e.g., the presentation may need substantial improvement.

As a product of science, a research paper, especially an important one, is intensely scrutinized even after publication. Sometimes, the experiments are repeated, and theories are re-derived. The scrutiny is much more critical and detailed than a philosopher's armchair speculation is because of the requirement of evaluation. There is an incentive for people to identify the weakness or flaw of an important result because this could be the easiest way for someone to become well known. Many young scientists are eager to find flaws in existing influential theories and observations. It is not an exaggeration that almost every Ph.D. student has made attempts of this nature. If a flaw is identified in an influential result, it would be exposed. In other words, if an important result remains standing and influential, it is agreed upon by the science community. For example, some of my own works were repeated or challenged. Admittedly, the scrutiny of less important works involves fewer scientists, and hence the chance of containing flaws is larger. Nevertheless, few people pay attention to the less important results. Their potential negative effects, if any, are less widespread. If one day, an unimportant result suddenly becomes important, say, because it is related to an important new result, the scrutiny into the old work would intensify. This would be the moment when the result will be brought under the microscope.

8.4 Can I Be a Scientist?

There is a wide range of abilities among scientists. A large fraction of them may be good at carrying out specific tasks, others are capable of making sound scientific decisions and opening up new research directions, and a few are leaders of a larger field with sound scientific understanding, judgments, and strategic scientific vision. Of course, all scientists have to go through the first stage where they are trained to become a scientist. Most of the capable and ambitious scientists want to move to the second stage, but a significant number of people may be attracted by other career opportunities while carrying with them what they have learned in scientific research to different career paths. This is understandable because science in most countries is traditionally not a particularly financially rewarding career while requiring extraordinary capabilities. This provides a natural filtering and selection process. After many years of practice, most people who decide to stay in science, as a career, are in the second stage.

The readership of this book is assumed to be the general public, who are interested in science and engineering, with a focus on those who are interested in learning more about the second stage of a scientist. After learning about the excitement as well as the frustration of being a scientist as described in this book, even if a person decides to do something else, they will think more like a scientist. In particular, they would ask more questions starting with "why" in contrast to "what" and "how". But be careful, in the non-science world, the word "why" can be intimidating to your boss!

In general, scientists only study problems that no one knows the answer to, problems that may appear to have no answer, or problems that no one knows about their nature. If people know the answer to a problem, it is an engineering or academic problem. Therefore, almost all problems in textbooks or exams are not currently active scientific problems because, with a few exceptions, the person who made the problems knows the answers. Also, *scientists* have to address real (not only idealized) problems. In contrast, those who possess both broad and systematic knowledge and can explain knowledge in a systematic way are categorized as *scholars*. Scholars may or may not be able to solve difficult real problems. They are more like a professor in a military academy who explains a war or battle but has not directed one in contrast to a General who has directed and fought wars or battles. A major difference is that the General is more likely to make mistakes but often wins the battles/wars (otherwise, he would not be a General) and the professor, who directs a battle on paper, can always tell you the best way to win a battle. Similarly, scientists may make more mistakes (before publishing a work) than scholars, as I discussed in earlier chapters. Of course, it is better if one is both a scholar and a scientist at the same time, but few people can be both because the two require a different set of talents and to succeed in either is extremely challenging already.

Scientists are a relatively small group of people in society who have chosen to do scientific research as their lifetime career. Many people participate in science projects and many students obtain scientific training or degrees. But they are not necessarily scientists because scientists, as a group, undergo a filtering process. Many people started in science, but they later pursue other career opportunities, e.g., Kuhn, or become assistants to scientists or teachers of science subjects.

Scientists may share some common personality traits. Almost all scientists agree that the driving force for choosing science as a career is their curiosity. Scientists' passion for science and their curiosity take over their life. What they enjoy most is the satisfaction after successfully solving a scientific problem. In the process, one can experience the excitement of realization of personal intelligence and value. If making money is a very high priority in one's life, the person is very unlikely to become a good scientist. Among scientists, we often do not discuss how to make money but joke about what to do next if one has made enough money — science is still *the* most interesting and challenging thing to do. There is no end of questions to answer and new knowledge to invent for the whole human beings; everything else would be boring after a while. Making money is a relatively low priority in life, although scientists do want to make sufficient money for a decent life.

Curiosity makes scientists more sensitive about things they encounter or hear. This is an important ability for successful scientists. Scientists do not stop to marvel at natural, social, and cultural wonders. They are also investigative, by trying to make sense of these phenomena. In this investigative process, connecting phenomena of different natures or knowledge from different sources is also an important ability to be a successful scientist. Of course, more often, they make mistakes when they apply their knowledge to fields that they have no expertise in.

For example, physicists have tried to apply the second law of thermodynamics to society and human beings. The second law of thermodynamics states that the entropy of a closed system cannot decrease where entropy is a measure of disorderliness. When entropy increases, the temperature tends to increase with it. These physicists explained human life as a process of increasing entropy — newborn babies have the smallest entropy. When people get old, the disorderliness, or high entropy, in the body makes organs malfunction. When the entropy is too high in a person, they die. Therefore, some predicted that the world would become so hot that everyone dies. If you have happened to hear about this idea, I will tell you that this explanation is false. This is because the human body is not a closed system. We breathe, eat and drink, *and* excrete wastes. If one thinks that the food we take in is more organized than the wastes we excrete, a lot of entropy can be removed from our bodies. One cannot simply apply the second law of thermodynamics for a closed system to a human body. Take the solar system as another example. It started as a cloud of gases and dust. Now, we have the Sun with planets circling it. Is a cloud of gases and dust more organized or is the system of a star with planets moving around it more organized? You have to agree that the latter is better organized and hence has smaller entropy. How come? Is the second law of thermodynamics wrong? The answer is that the solar system is not a closed system. There is radiation, which removes the heat from the system so that the entropy of the solar system actually decreases. But mistakes like these do not stop scientists from trying again. These days, entropy has become a hot topic among researchers in economics and sociology.

Another fundamental personality trait that distinguishes a scientist from others is critical and self-critical thinking. Scientists do not take any explanation or result as given, as stated by Descartes. They have to constantly challenge authorities' ideas and current knowledge. As a professional habit, they often question authorities outside of their specialized fields, like what I am doing now when writing this book, and can be too cynical or annoying sometimes. They are also very rational. They think that most phenomena are connected in some ways and may have causal relations. Although the details may be unknown, they can be worked out if there is enough time and information. Scientists do not take nonsense as proof, such as analogy and pragmatic argument as discussed in Section 7.2. Scientific reasoning sets the ground rules for scientific exchanges and consensus-building. Many philosophers of science assume that there is a scientific method. To them, the main question is how to find the right one. They have been building their theories completely upon the misunderstanding of science and scientists.

Often, I have been asked by parents about future plans for their children. I have found a behavior that may be characteristic: When a child was young and playing with toys, did the child tend to break the toys, e.g., a kaleidoscope, or not? If a child often breaks toys, such as by tearing them apart, the child is most likely curious with a tendency for being a scientist. If the child tended to improve the toys, they may have a tendency for engineering. However, this does not mean that scientists are more destructive and engineers are more constructive. Because of curiosity, a child may try to open a toy to see how it works. Other kids may focus on the toy's functionalities for doing different things or doing things differently. Both can be creative. On the other hand, if a child takes the toy they are given and plays with it, the child may be more operational, as a good manager. I am not proposing a theory or model here to predict a child's future career development but to point out some possible correlations. I was one who broke any toy in half a day. Children can change as they grow. Do not label them when they are still young and have not been exposed enough to the world. Parents should let children choose their life path, but pay attention to their intellectual development.

To be a scientist, one has to be rational, investigative, critical, and self-critical, not only curious. Intelligence and organized knowledge are also needed to sort out a complicatedly entangled network of problems. As we discussed before, creativity may have been overrated by the general public. When one encounters a problem without time constraints, good ideas will emerge; whether one can recognize those that have potential leading to the solution of a problem in their lifetime may be more important. There are a few secondary personal traits that may affect the success of a scientist and their achievements.

Deep versus superficial: It is generally true that some people tend to be satisfied relatively easily with an answer while others keep asking why. A salesman's understanding of an idea cannot be sufficient for a scientific problem. This personal trait can be recognized easily in college and graduate studies. Some students feel they easily understand everything they have been taught while other students find it agonizing to understand the intrinsic connection among different pieces of knowledge. In science, we definitely need to think deeply.

Stubborn versus flexible: At first glance, it may seem clear that flexibility is good and stubbornness is bad. But in science, this may not be the case. As we discussed before, any new idea in science will have to go through a falsification process. The more important the idea is, the more intense the falsification process would be. Imagine that everyone around you is criticizing your idea. Can you still think that you are right and everyone else is wrong? Some of the people who are against your idea may be highly respected scientists. Can you still insist that you are correct? Being stubborn in general is a virtue for a scientist as one needs to defend one's ideas. However, one has to carefully listen to and examine each of the criticisms. One cannot simply repeat one's own argument but has to study the negative arguments and explain why they are flawed or irrelevant. Self-criticism is essential in the process. One, while being stubborn, has to be ready to modify the idea in order to accommodate reasonable criticisms.

Broad and systematic versus penetrating: Some scientists master broad and systematic knowledge and are able to explain each piece of knowledge clearly. Other scientists are able to penetrate deep into a complicated problem and find a solution. However, the second type may not appear to be as knowledgeable as the first type. But often, the first type may not be able to solve complicated problems. As discussed above, we often refer to the first type as scholars and the second type as scientists. Of course, we want to be a scientist and a scholar at the same time. In the case where one does not have both traits, in science, one may want to find a collaborator with complementary abilities. In a science community, there are many lifelong collaborators with matching abilities.

Open and resourceful versus closed: In science, we see many people agonize over a problem for a long time without seeking help. This may be because of the desire to protect original and creative ideas, or the person is a deep thinker, as discussed above, especially in old times. Nowadays, information is widely available, but knowledge is highly specialized. Some of the problems could be trivial to others. An open style in research may be much more efficient, especially if the problem is not the core of the new idea of the investigation, such as how to use scientific software. Although there were few instances in my career where other scientists took my ideas and published them before me, I still feel that scientists often benefit more from open-style discussion. Open style often wins more respect.

Popper once said that good or great scientists can come up with imaginative, creative, and risky ideas. With hard-headed willingness in a subject, imaginative ideas can progress to rigorous critical testing. I found this a good portrait of a scientist.

8.5 Types of Scientific Research

There are many types of scientific research, each requiring a special set of talent and skills. Scientists would enjoy the process, as well as the challenges, differently. I discuss these different types of sciences according to Kuhn's theory that we learned in Chapters 5 and 6.

The most challenging type of research is to connect and explain several observations when there is no scientific paradigm or guiding theory to connect them. This is an effort to propose new laws such as done by Newton whose laws connect and explain the free-fall experiment, Kepler's laws, lunar tides, and many other observations, or by Maxwell who connected electricity and magnetism. I envision that in science disciplines that do not yet have firm paradigms developed, such as biology, psychology, medical science, and economics, there are great opportunities for ambitious and talented scientists to do this type of research.

In disciplines that already have major well-established paradigms, a new idea derived from one discipline may be applied to problems in a different discipline. Or, on an even smaller scale, one may apply a new idea developed in one subdiscipline to problems in other subdisciplines. Because the observational techniques and methods are becoming more and more complicated and specialized, the barrier between two disciplines or even between two subdisciplines has become increasingly difficult to penetrate. Research of this type requires substantial technical breadth as well as depth. Leading scientists in each of the disciplines mostly have conducted research of this nature in their field. For example, in space physics, connecting multiple observations from different subdisciplines within the discipline can be very challenging and few scientists are able to successfully achieve it.

Most scientists conduct their research under a known scientific paradigm, as what Kuhn described as normal science. However, the research is not as dismal as what was portrayed in some theories of philosophy of science. It is still very challenging because observations, most often, are not consistent with predictions, where inconsistencies can be caused by various potential causes, especially by incomplete information. Scientists have to be able to understand the whole network of knowledge that is concerned by Holism. They have to quantify and explain some of the regularly observed properties or features and abnormal observations as well as the convoluted theoretical models. In the process of resolving these inconsistencies, a scientist can be challenged intellectually, and hence they can enjoy the satisfaction of achievement after the problem is resolved while also contributing their intelligence and creativity to the human understanding of the world. The satisfaction of achievement cannot be measured by money or fame, though if money and fame come with a scientific result, who would refuse? This effort can be glorious and rewarding, e.g., the discovery of Neptune via Uranus's orbital aberration. Creativity is needed in the process. However, it is not as magical as imagined by the general public. Most experienced scientists have moments in their career to enjoy the satisfactory feeling of solving a seemingly unsolvable problem. However, the scientific impact of that moment depends on the importance of the problem being solved. One has to be working on a potentially important problem; this is where philosophy of science can help, as will be discussed in Chapter 11. However, after reading this book, everyone would swarm to subjects of high impact, thus making the competition in the subjects keener and the chances to be the first to solve the problem smaller. Nevertheless, if you happen to be prepared for a subject, go for it and do not worry about the competition! If the problem is truly important, number 2 or number 3 will also be recognized in the science community although only the number 1 may be able to draw attention from the general public.

Scientific breakthroughs often result from new observational techniques and developments. Information collection includes designing and carrying out experiments and tests. A better thought through experiment is always essential to the success of a study. Therefore, although debatable, careful identification of essential information or measurables with feasible measurement techniques and instrumentation is important to determine whether a project will succeed or not. For example, in space physics, it often takes decades to debate about the scientific objectives and resolutions of measurable parameters to decide the instrument's capabilities and designs for a space mission. Many students obtain their first experience of science by participating and conducting experiments in this type of research to understand a phenomenon under some guiding ideas. Sometimes, the guiding idea is clear (when the adviser is an experienced scientist), but at other times, it is vague (when the adviser is not an experienced scientist or the subject itself is not well developed). In the latter case, collecting information about a phenomenon with some vague controlling factors may not be a good experience for a student to learn about science as students may develop misunderstandings about true science.

Data analyses are an area of research that may need more "creativity" because the rules are often less rigid. Scientists and maybe students of various backgrounds can participate in them. Pieces of information can be organized in various manners and correlated with a wide range of possible parameters. The entry barrier is usually lower in this type of research. However, uncontrolled "creativity" can produce flawed conclusions. For example, I just learned that some of the erroneous predictions in the 2020 presidential election were attributed to the weight function introduced to the original data that was supposed to correct sampling biases. The weight function may actually adjust a result to what the analyst likes. Real science, fundamentally, does not allow this. If a scientist wants to introduce a weight function or correction to the data, the uncertainty in the result should be correspondingly *increased* and not reduced. Therefore, the "creative" data analyses used in some model predictions were fundamentally flawed because the error bars were not increased accordingly. The "creativity" and "creative ideas" presented in posters as the light bulb in a human head, in the context of science, to me, mean something suspicious because this is where flaws are introduced to the results. In science, great ideas come from a long period of deep thinking. One would try out every obviously or distantly related possibility and remember every one of the failed ideas. When an idea truly carries a plausible solution, one is able to recognize it and everyone else would think of it as a lightning flash. Without those failed attempts, there is no creativity in science.

When observation is consistent with theoretical predictions, unless it is the first confirmation of the theory, straightforward explanations of the observation have little scientific value. AN explanation based on intuition, instead of scientific understanding, does not qualify as science. As discussed previously, most often, scientific research is needed when there are inconsistencies between an observation and a theory or when there are multiple theories that can explain an observation. In the former case, the theory may need modifications or the observation needs improvement. In the latter case, theoretical models need further development of specific details in order to differentiate among them and the observation must be more specific on these observed features.

Nowadays, computer simulations are widely used to provide detailed features as well as quantitative predictions. Here, I should point out that artificial intelligence and neural network, except in few cases, have not shown potential in providing scientific understanding or knowledge-understanding. Therefore, although it is useful in engineering, business, and certain applications and may provide good predictions, their predictions are *not* based on scientific reasoning until they can provide knowledge-understanding.

8.6 Different Perspectives Between Scientists and Philosophers

First, most conventional theories of philosophy of science and the scientific method were developed based on current understanding and judgment of historical events or episodes in science. Scientists, on the other hand, may be currently involved in an event that is similar but may not be on a scale as grand as that of Copernicus, Newton, or Einstein. The scientists do not have as complete information as historians or philosophers do centuries after an event and do not know the future evolution or development of the event. They also do not have the luxury of "armchair speculations", as most philosophers do, because they have to make judgments and decisions based on the available information at the time to guide their research. Some decisions can be critical in determining the future of their scientific career. For example, at the time when Newton's universal law of gravitation was first invented, a scientist could be very puzzled by the theory that states gravity is inversely proportional to the square of the distance between two objects. If the distance between two pieces of earth goes to zero, shouldn't the force go to infinity? Shouldn't the whole Earth clump into a small ball? Of course, we now know why this does not happen — because of the pressure gradient force that tends to push the matter away from the center of the Earth. At the time, a scientist could think Newton was crazy. The rationalization developed a long time after the event has little value to a scientist at that moment.

Second, in science, there are always two or more competing ideas. More often, both sides are scientists who are all rational, but their work is based on different paradigms or approaches. Their ideas each appear rational to themselves. Therefore, scientists seldom make decisions between rational and irrational. Scientists need to know which rationalization presented is *more likely* to be correct. Without knowing the future and only having incomplete information at the time, scientists may find that the history and rationalization discussed by some conventional theories of philosophy of science have little value at best and that the rationalization developed was based on interpretation or misinterpretation of the events, and results may have distorted history. Note that the logic for many of these theories is abductive: One makes a hypothesis, distorts the facts to fit the theory, and then declares the success of the theory. For example, following the rationalized theory of philosophy of science, Galileo would have falsified Copernicus's theory. Who would believe such a theory?

Third, most conventional theories have been developed based on publications and historical records. Unpublished ideas are not accounted for. This, on the surface, appears to be unbiased and objective as the theories are based on hard "evidence".[2] However, as described in this book, instead

---

[2]An interesting similar example is that some historians developed a theory according to which people's lives were better during periods of civil wars than during peaceful times. During civil war, the central government was overthrown by warlords or rebels who collected money directly from citizens without record. Namely, according to historical

of justifying their hypothesis, scientists spend most of their time rejecting most of their creative ideas. Uncountable unsuccessful investigations and attempts have not been published and there is no record of the learning process and reasoning for rejected creative ideas. Many scientists can learn a great deal from these unsuccessful studies and others are frustrated and struggle and have to leave science as the career they had dreamed about when they were young. Although those who have left would not be able to write about the lessons learned, their colleagues, friends, and referees of their work carry the memory of these failed attempts. But to many conventional theories of philosophy of science, these never happened. The conclusion of the famous example of Russell's chicken farmer problem may be valid if there are generation gaps that make knowledge accumulation impossible as we discussed in Section 4.3.

Fourth, most conventional theories of philosophy of science have been a natural extension of general epistemology in philosophy. In epistemology, "knowledge" is general. Commonly used in the studies are the survival instincts of monkeys, dogs, mice, or chickens. These instincts are fundamentally different from the logic and reasoning used in science by scientists. Furthermore, most examples based on which many theories of philosophy of science were developed are what is referred to as "knowledge-that", a piece of knowledge in isolation from other pieces. In more advanced examples from psychology or neuroscience, such as how many randomly arranged characters, pictures, or numbers one can correctly remember and retrieve from their memory in a given period of time, the issues tested are "knowledge-that". These are still irrelevant to the process of invention of knowledge-understanding. On the other hand, modern scientists are a group of highly trained professionals, different from chickens, dogs, monkeys, mice, and even ordinary folks. Using such examples to draw conclusions about how scientists should think is very problematic and, at best, irrelevant or inappropriate. It is not only not useful but also misleading or confusing in discussions of philosophy of science. Science is not about how to learn knowledge, but to invent new knowledge-understanding which looks for connections and relationships among

---

records, the government collected no tax during civil wars. This fact has been interpreted as people not needing to pay tax and hence had a better life during civil wars, a flawed conclusion based on apparent unbiased fact.

isolated pieces of knowledge-that or knowledge-understanding. To a scientist, with the understanding of physics, the question of why we see the Sun moving crossing the sky every day is not a simple philosophical issue of using inductive inference to predict whether the Sun will rise tomorrow. On the other hand, scientists are trying to solve specific science problems in their corresponding disciplines. Very often, when a scientist comes up with a new idea, they want to know whether it could be wrong. Simple logic, as discussed in conventional theories of philosophy of science, is not sufficient to address issues like this.

These four fundamental differences in motivation, subjects of interest, and experience may distinguish the perspectives of scientists from those of philosophers of science. These scientists have failed and succeeded many times in the process of scientific investigations and inventing new knowledge. They also witnessed the failures as well as the successes of their colleagues and the progress of the field. They have learned a great deal about philosophy of science consciously or subconsciously from these events and experiences. They have to make their judgments based on only limited or incomplete information at the time of the emergence of a scientific problem. They do not know how history will play out in the future. However, they do care about how their ideas will play out at least before their death. Scientists who love philosophy of science are trying to develop theories that concern scientists over diverse science disciplines based on their scientific research experience and philosophical concepts.

There is something that scientists and philosophers have in common that differs from many other professions. Both professions do not take anything as given and we do not believe anything without evidence and reasoning although evidence may be influenced by personal bias. Authorities, prestige, or fame do not carry additional credibility, although, sad enough, to the general public, these do carry substantial additional credibility.

8.7 Successes and Failures of Conventional Theories of Philosophy of Science

Throughout this book so far, you may have seen many of my criticisms toward the existing theories of philosophy of science. However, these previous studies have flushed out many important issues. Some of the ideas and thinking of conventional philosophy of science are very insightful and useful: the importance of observation according to empiricism and reasoning according to rationalism. We learned from Descartes to think critically, to "divide and conquer", to think in an orderly fashion starting from the simplest and easiest to understand and from important to less important, and to make enumerations complete. We recognized potential problems using inductive inference and explanation. We learned from Popper the importance of falsification, the problems with "scientific methods", and the importance of risk-taking. Falsification theory catches a crucial part of scientific activities — dialectical reasoning. Kuhn brought us the concepts of the paradigm, normal science, and revolutionary science. Identification of the existence of a paradigm, no doubt, is one of the most insightful contributions to philosophy of science.

An important contribution of philosophy of science is to theorize the separation of science from metaphysics. Debates over metaphysical issues can be extremely confrontational and unproductive in science. For example, to most science and applied science disciplines, what causes gravity may be considered a metaphysical question although it is a science question in relativity studies. Without the issue of the cause of gravity, one can potentially dismiss the metaphysical debate about the universal law of gravity. To most scientific disciplines, the law is extremely valuable and the metaphysical concern is not an issue.

The framework of conventional philosophy of science, however, has failed massively. Over the last hundred years, philosophy of science has taken two major initiatives: finding a scientific method and logically reconstructing science and the history of science. As I have discussed throughout this book, the promises have not been fulfilled and more confusion has been created in society.

An important issue that many conventional theories of philosophy of science have overlooked is the difference between observation and evidence in science. In science, evidence is often *not* the observation itself and is *not* as unambiguous as a murderer's weapon in a courtroom. Evidence is often an interpretation of observation based on a specific rationalization. For example, when technical details are ignored, assuming that the experimentalists can handle the issues properly, the fact that the ball dropped from the Leaning Tower of Pisa and hit the ground right beneath the dropping point is an unambiguous observation or empirical "truth". However, this fact can be interpreted as evidence for either the Earth not moving (before Galileo) or moving (after Galileo). On the other hand, if one drops an iron ball and a feather ball of the same size and shape, the iron ball would reach the ground before the feather ball. This fact, if presented before Galileo, can be interpreted as evidence that a heavier object experiences greater gravitation acceleration than a lighter object. (We now explain that the feather ball experiences the same gravity but has more air resistance than the iron ball so that it falls more slowly). To scientists, if the evidence "depends on" rationalization, it cannot be used as the basis of rationalization; otherwise, it is circular logic. Science is not simply the problem of rational versus irrational nor is it about being either based on evidence or not based on evidence. Science has to find the correct rationalization to explain an observation without knowing what the correct rationalization should be when information is partial. In this regard, the conventional philosophy of science has, so far, completely missed the point, by having focused on things such as linguistic rules and forms of logic.

Conventional philosophy of science failed to deliver what it set to provide: a strong "prescriptive moral", a universal logic for induction of confirmation, a universal scientific method, and a universal paradigm/ scheme/framework of science. It debated about whether the rules are good and reasonable and whether a theory of science should be "descriptive" or "normative", i.e., the way things *are* or the way things *should be*, respectively. The normative issue is important to philosophy, sociology, or politics but generally carries no weight in determining the correctness of a scientific idea because nature does not care about how humans think it should or should not be. The fundamental flaw in this line of thought is that it assumes that what is good, reasonable, or "should be" can be found. Science concerns the problem of knowledge and not the problem of conduct. There is no choice in the problem of knowledge.

Conventional philosophy of science has been carried away by the suspicion of induction in science, a notion directly inherited from epistemology, resulting in fundamental flaws in the overall logic in many theories. The theories correctly recognize the potential fallacy of inductive logic, but the examples used concern only knowledge-that but science concerns mostly knowledge-understanding. Unfortunately, the theories conclude according to the flawed reasoning that induction cannot be used in science. This was the first fundamental mistake in discussions of logic. As a result, the science theory has to rely only on deductive logic. Since deductive logic is from general to specific, if a theory is based on deduction alone, it has to cover everything from start, according to the logic. Therefore, a theory must be universal and exclusive. This seems self-consistent, but science is more based on limited observation which cannot provide a universally true starting point without generalization. Without induction, there is no way to reason from specific to general; pure deductive logic is handicapped. The second flaw is that many theories consider abductive logic a valid scientific logic without recognizing that it is deeply flawed logic. The third mistake is that most theories completely ignore the dialectic logic introduced by the founding fathers of philosophy! The massive confusion about the logic in science eventually misled many conventional theories of philosophy of science.

Producing a scientific method was a valid investigation. If found, the method can streamline scientific research so that science can be done by people with less scientific training, like in a production line. However, it failed to deliver a universal foolproof one because the proposers overlooked the diversity of science problems and scientists. Few experienced scientists agree on any scientific method. They found the methods discussed irrelevant to the reality of scientific research. I have shown that when following the popular scientific method, one is more likely to either do trivial investigations or derive a wrong conclusion. The development of scientific methods was misguided by the suspicion of inductive logic. The proposers did not understand how scientists think and how science is conducted. The common problem each individual scientist encounters is that there are too many creative ideas and possible approaches to choose from. One is able to test only some of them. When testing them, one only encounters more problems that require many more creative ideas. Eventually, only a small percent (according to Einstein) of creative ideas bears fruit. In such a process, no "scientific method" would be able to provide a guide to rejecting most of one's own creative ideas, let alone a recipe for everyone. Some may argue that a good chef does not follow a recipe; that is only for those who do not know how to cook. This may be true, but most people are not good chefs. The worst situation might be that a good chef could be fired because of not following the recipe given by a manager who does not do any cooking.

Rationalization of science, and reconstruction of the history of science, was a flawed concept from the beginning because it assumes that the re-constructer knows the right rationalization, but history does not proceed according to any rationalization. When theories of philosophy of science could not explain the history of science, some theories suggested rewriting the history of science in order to fit the rationalization provided by the theories (Lakatos, 1970). However, the theories have not explained how to know their own rationalization is NOT wrong. The implicit assumption for rewriting the history of science is that the proposers know the path to truth. This is a fundamentally flawed assumption, a circular logic, in all the related investigations. The theories do not recognize that science may take different processes/paths through their experimentations and interpretations. The idea that one can rationalize the history of science is a fabulous dream. In the history of science, examples cannot be used to *prove* any theory but may be used to *illustrate* how a theory may work if it has any possibility. Similarly, counterexamples can definitely be found against any theory. One should not expect a universal, exclusive, and absolute theory to work. Exceptions should be allowed for a theory. Then, one would find that there are too many exceptions that essentially diminish the meaning of a universal theory.

Science, on the other hand, seems to be able to defy all conventional theories of philosophy of science by developing new technologies and methods, such as computers, smartphones, and satellites, successfully based on the progression of our knowledge-understanding. Some of our knowledge-understanding is definitely correct or true, not as Popper assumed.

The root of these massive failures of the conventional theories of philosophy of science may be the extremism presented in the debates of the field. Note that extremism is required by deductive logic. According to this radical idea, every theory has to be universally true and exclusive to all other alternatives, e.g., either justificationism or falsificationism. No scientist would agree with such a radical idea without alternatives or constraints by observation and evaluation. According to the founders of philosophy of science, the answer is simple: A scientist has only one choice based on one's philosophical belief, either empiricism or rationalism. When a scientist encounters a difference between observation and theory, if the scientist is an empiricist, the theory is wrong or if the scientist is a rationalist, the observation is wrong. Although modern pluralism does not insist on picking one extreme, it does not explain how to make the right decision among alternatives.

In many theories of philosophy of science, people "armchair speculate" all possibilities without constraint. In science, when adding the element of evaluation to the requirement, most irrelevant possibilities can be consciously removed after debates in a peer-review process, or caveats can be imposed to the conclusions. Most armchair speculations become irrelevant. In science, the answer could depend on whether a set of assumptions of a theory is applicable or whether the range of uncertainty and limitations of observation or instrumentation is sufficient to resolve the concerning issue. Any observational result has a range of possible interpretations. Using empiricism and rationalism to categorize philosophy of science misses the point. Instead, one of the tasks of science is to gradually reduce the uncertainty in both theory and observation. This is where "evaluation" plays a key role. Holism, according to which science is unable to identify flaws in a network of involved knowledge, fundamentally misinterpreted the situation.

Using Feynman's analogy, the conventional theories of the scientific method or philosophy of science to scientists are like ornithologists commenting on birds. They categorize the birds from the viewpoint of an outsider. They do not understand the language of the birds, but they assume that they are much more intelligent than the birds and know why the birds are doing a certain thing. Why should birds change their behaviors according to ornithologists who do not understand the reasons for the birds' actions? The objective of this book is to fundamentally change the situation and the framework of philosophy of science.

Questions for Thinking

1. If you hear from the news that someone invented a new idea to treat cancers, is this science or knowledge? Please provide your reasons.

2. If you read an article that says the continents were originally connected but are drifting away, is this knowledge or science? Please provide your reasons.

3. Can you name where the subject you have chosen to study is in science? Which branch? Which discipline? Which subdiscipline?

Chapter 9

Theory of Science

9.1 The Overall Structure of Science

The objective of science is to invent new knowledge and, at the same time, to filter out false knowledge. Science comprises three key elements: observation, conceptualization, and evaluation; these are connected by reasoning. Observation and conceptualization are the two fundamental sources of knowledge; evaluation distinguishes science from all other forms of knowledge inquiries, such as philosophy[1] and history.[2] Inferences can be made following an evaluation. However, this raises a philosophical question: What if unflawed inferences from the two sources do not match? When this occurs, in general, scientists may have to invent various ways, e.g., by modifying the theories or tinkering with the numbers, to show that the inference and observation are consistent. This may be possible when we only have partial information so that both the inferences and interpretations of the observation have uncertainty which provides room for

---

[1] Ontological debates and many of the most fundamental questions in philosophy cannot conclude after thousands of years. Conventional philosophy of science is still debating between empiricism, rationalism, and other various "-isms" without apparent progress, as discussed in Chapters 7 and 8. A significant cause of this situation is that they do not have an evaluation system to control these debates. Without quantitative control, one is unable to remove a remote possibility, such as the flap of butterfly wings, when discussing a significant magnitude phenomenon, such as earthquake.

[2] Due to inherently considerable uncertainty in the interpretations, especially the motive of each player, history has little probability of being quantified and, hence, to become a science.

manipulations. This uncertainty would produce another problem — a single observation accommodates multiple inferences. The main purpose of science is to ensure that observations, conceptualizations, and evaluations are derived rigorously, which tends to reduce the uncertainty in the inferences and interpretations. Eventually, the uncertainty becomes so small that theories can be differentiated. A combination of the three key elements with *scientific* reasoning addresses the question of how science can get things right. Therefore, all possible inferences and comparisons, as well as the methods used for comparison, must be (eventually quantitatively) evaluated.

We have discussed various forms of reasoning in Section 7.2, though only four forms of reasoning are generally accepted in science. These forms are scientific deductive, inductive, and dialectical reasoning plus Occam's razor. However, each of these reasoning forms only serves certain functions or purposes in inventing new knowledge and each has its limitations and possible fallacies. These limitations and potential fallacies are widely known, consciously or subconsciously, in science communities, especially by experienced scientists, though these limitations may not be well known to the general public nor to most philosophers of science. In school, there are no systematic teachings of scientific reasoning beyond the logic taught by logicians. Therefore, in this chapter, I provide a comprehensive discussion of each form of scientific reasoning. After going through the discussion, the questions of "how to get it right" and "how to know I am not wrong" are addressed in principle. As the second question is most important but more difficult to address, we will be further discussing this in Chapter 11.

Without a standard or a commonly accepted scientific method, how do scientists settle their disagreements? Compared to politics, in which a conflict could result in the loss of lives or scandals, science is relatively more civil and peaceful. This should have been one of the most important issues for philosophy of science and/or sociology of science to recognize, discuss, investigate, and answer. As discussed in Chapter 8, science is built based on honors principles and is a disciplined endeavor. Although intellectually each scientist may be somewhat anarchy, they belong to their scientific communities, where they must behave professionally in scientific activities. Their ideas have to be presented following scientific reasoning; otherwise, they are considered flawed. The organization of professional societies is based on honors principles with democratic elements. No one is above others. The organization, in addition to providing forums for scientific exchanges and a learning environment, regulates and monitors the peer-review process for scientific publications and proposals for new scientific projects. Such a system is able to promote an individual's creativity while allowing opposite ideas to coexist in civilized manners and preventing unproductive behaviors from being out of control. No one can guarantee that this system would always work perfectly, and anomalies do occasionally occur. However, the scale, intensity, and frequency of these anomalies, in general, are very limited in most scientific fields, especially compared to, e.g., politics.

The conventional theories of philosophy of science introduced in Chapters 4–6 and to be discussed in Chapter 10 do not really touch upon the central questions scientists face: How to get it right? And how to know I am not fooled by myself? For example, a simple acceptance of either empiricism or rationalism does not help much because all scientists believe in each to a certain level. They all base most of their ideas on observations, but whose idea is *correct* or *more correct*? Without knowing how to make judgments in relation to this question, all theories of philosophy of science would have little relevance to scientists. This is the problem we will address in this chapter. We start by first discussing the four forms of accepted scientific reasoning and then follow by sketching a guideline of how scientific judgments can be made.

9.2 Scientific Deductive Reasoning

The deductive reasoning used in science is somewhat different from the deductive logic commonly used in philosophy of science which emphasizes the "form", as discussed in Section 7.2. In science, it is mostly used to derive an understanding of specific problems based on general knowledge. This reasoning process starts with commonly accepted laws or assumptions and is followed by straightforward deductions. When including the evaluation requirement in science, most often, scientific deductive reasoning can be carried out by mathematical derivations and/or numerical solutions from a set of governing equations. Deductive reasoning is forwardly deterministic and truth-preserving similar to a plumbing pipe: if the input is right, the outcome is right!

Deductive Reasoning and Mathematics: People often directly relate science with math and assume doing science is doing math. This is because mathematics (not the law of syllogisms!) is the most important form of deductive reasoning in science. This is a little bit magical to a scientist. When they solve a very complicated set of partial differential equations — how could a scientific problem become a mathematics or computer programing problem? Can this change introduce errors?

One has to first be able to derive mathematics correctly; this is easier said than done. For example, in theoretical physics, part of the basic training is to do a 100-page derivation without an error. In space physics, we often deal with a minimum set of 13 first-principle-based partial differential equations covering information from the surface of the Sun to the ionosphere. How do we know the calculation will correctly give us what we want? In a common case when the calculation result is not the same as the observation, should we blame the theory, the calculation, the observation, the data processing technique, or the instrument itself, the question that frustrated holists? Scientific deductive reasoning tells us that if our theory is purely mathematical and we do not make mistakes in derivation, we can trust the math. If something is wrong in the theory, the error can be caused by anything else but not the mathematical procedures as long as the conditions for the procedures are satisfied. This is the power of mathematics!

Many scientific laws are conceptualized based on observations and are written in the form of mathematical equations. Applied sciences and engineering develop models based on these laws so that engineers can solve and compare the solutions of these equations numerically with different effects under various corresponding initial and boundary conditions. A fundamental assumption in science is that if deductive reasoning is based on mathematics, there is no leakage and contamination in mathematical procedures. Leakage can result in a loss of reasonable possibilities, while contamination can introduce false possibilities. It is true that mathematics provides a strict chain of deductive logic. However, it has worked "amazingly" perfectly, so I have to question it because mathematical procedures were developed independently of scientific laws and their mathematical forms. A scientific law "happened" to be written in a particular mathematical form but why does the mathematical procedure to manipulate the form have any relationship with the scientific problem we are studying? When a mathematical result is inconsistent with observation, can mathematical procedures introduce leakage or contamination in the process? Why not? This is the most puzzling question to me concerning philosophy of science even after I have developed the theory of science outlined in this book.

In scientific research, if solutions are inconsistent with the observation/expectations, we can question the equations used, the approximations made, the terms included, as well as the assumed initial and boundary conditions, but not the mathematical procedures! Why not? This is very strange! Nobel Prize Laureate Eugene Wigner once said, "the enormous usefulness of mathematics in the natural sciences is something bordering on the mysterious and there is no rational explanation for it". Relationships between science (other than math) and mathematics need to be understood better in philosophy of science although much effort has been made.

Unfortunately, only very limited cases can be solved analytically. Most often, the equation set has to be solved numerically using a computer algorithm. If the equation set is time-dependent, the solution would change with time and is often referred to as a computer simulation. In the analogy of a pure scientific deductive reasoning to a pipe's output, the mathematical derivation, like a properly working pipe, would be a well-sealed or isolated system. The numerical solution may not be perfect just as the pipe may be permeable. Therefore, it is possible for the pipe to produce some leakage and/or introduce some contamination. Depending on the numerical algorithm used, leakage and contamination can be either unimportant or substantial relative to the required accuracy of the problem. Therefore, possible fallacies or invalid reasoning can be introduced when a computer simulation is involved, even using the strict scientific deductive reasoning of mathematics. If the deduction is not mathematical, the output is not controlled quantitatively.

Reasoning Flow: Scientific deductive reasoning cannot go in a backward direction, from a known result to infer the cause of it, or it is abductive reasoning which is flawed. Abductive reasoning is a commonly invoked flawed reasoning when one proves a hypothesis that is used as the start point of the reasoning. Deductive reasoning within a scientific problem can be very complicated. Sometimes, it can eventually return to its starting point, i.e., the logic becomes circular and is flawed, such as in abductive reasoning. In mathematics, circular logic occurs when the number of equations is less than the number of unknowns. In this case, some of the unknowns are represented by other unknowns. Additional information is needed in order to solve the problem.

One can rarely solve a real scientific problem based on a single set of laws because someone else would have solved it. More often, it may involve several fundamentally different processes, where each is governed by a specific set of laws. There may be a need for some assembling work of pipes, i.e., there are connections of pipes. We all know that the joints of the pipes are where most problems occur. Therefore, carefully checking where different ideas or approaches meet is essential to any scientific investigation.

Figure 9.1 illustrates the process of conceptualization or development of our knowledge. The objective of the process is to make connections among knowledge-that and develop conceptualization and understanding based on available observation and knowledge; this process includes inventing new knowledge. In the figure, two phenomena are in the lower right and the upper left of each panel. Each phenomenon may refer to a group of phenomena, e.g., they may be described by a general law, but the two groups of phenomena may not be obviously connected, for example, free-fall and planetary motions, or an asteroid hitting the Earth and extinction of dinosaurs. Science is to find the underlying connections. The two may be connected in various ways, each of which can be considered a conceptualization or understanding.

The causality of reasoning is indicated by double lines with arrowheads. In principle, the causality of scientific deductive reasoning is forward, i.e., like water in pipes flowing from an input to an output, as shown as the vertical lines of reasoning from top to bottom in the left panel of Figure 9.1. Each line of deductive reasoning may, in fact, be based on a paradigm. Mathematically, the solution of an equation is the result and the input to the equation is the cause. Sometimes, there are multiple variables in an equation, for example, Newton's second law of motion has both the force and position of a body as variables.

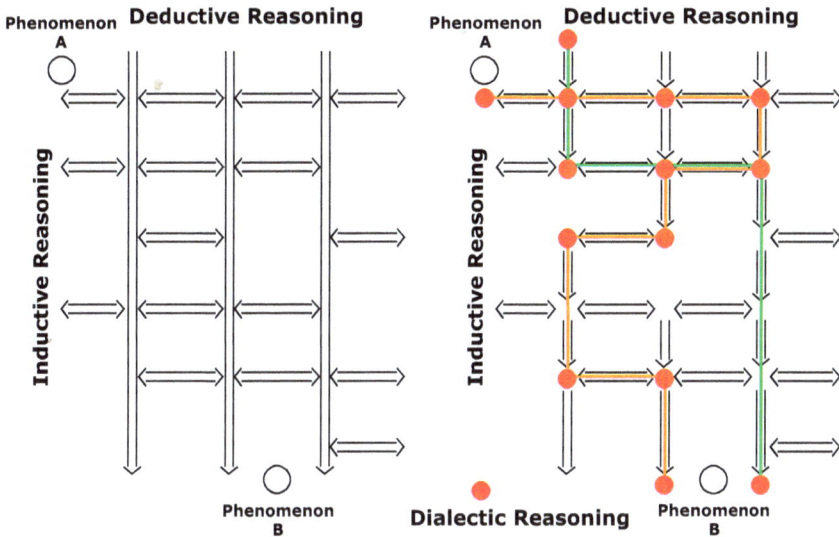

Figure 9.1. Illustration of the Theory of Scientific Reasoning. The plane describes human cognition with two phenomena or groups of related observations on the upper left and the lower right of each panel. The vertical reasoning flow is deductive, such as mathematics, which is forward in causal relation. The horizontal reasoning flow is inductive, such as a result from a correlative analysis between two observations/variables, which cannot determine the causal relation. Two reasonings/understandings/conceptualizations are shown as green and brown lines in the right panel. At each connecting point, dialectical reasoning, indicated by red dots on the right panel, is used to eliminate alternatives.

As discussed in Section 7.5, if there is no time derivative in a governing equation, such as in a quasi-steady-state system, such as ideal gas law and Ohm's law, no causal relationship within the system can be described. The cause has to be external changes. If the governing equation is dynamic, i.e., by including a time derivative (e.g., acceleration in the case of Newton's second law of motion), the net effect corresponding to the forces is the cause of the variables (speed and position) of the time derivatives.

Potential Problem with Deductive Reasoning: The most important problem is that deductive reasoning itself cannot provide a starting point. When one puts wrong things into the pipe, i.e., the input is incorrect or not optimal, the system cannot help to sort out the flaws and results in a situation called *garbage-in-garbage-out*. Without a justified starting point, the reasoning is "infinitely regressive"; people can question the start point and chase it backward infinitely as discussed in Section 4.2.

To address the problem, the starting point has to be "self-evident". However, when you think something self-evident, someone else would have thought the same and have tested it. The result can be either a law of nature or is unlikely true. Therefore, the self-evident argument is used less often in science except in underdeveloped fields. A self-evident starting point should have been described by the laws of nature. When a law is invoked, *applicability* and *relevance* are the two issues that will have to be carefully scrutinized.

In a field of research where laws of nature do not exist, one can invent one based on observation or experiments via inductive reasoning. However, inventing a law is not a trivial effort; a large amount of observation and experiments under various conditions and in different processes may be needed. This is because a law of nature requires *universality*. Observations or experiments for a single process under one condition will not be considered a law of nature. If inventing a new law is not an option, one may directly invoke robust observational facts under the conditions of interest. However, the generalization of such an observation can result in potential fallacies. As we have learned, the interpretation of a fact usually involves or is based on inductive reasoning and may not be unique. If the interpretation is not unique, everything is questionable.

Since a law of nature or interpretation of an observational fact has to involve inductive reasoning, deductive reasoning by itself cannot derive a scientific result. The idea that philosophy of science can develop a theory of science solely based on deductive reasoning is a dream.

9.3 Scientific Inductive Reasoning

Generalization and Conceptualization: Generalizations of limited observations are based on inductive reasoning in science similar to the inductive logic discussed in conventional theories of philosophy of science. However, in order to avoid the fallacy as discussed in conventional theories of philosophy of science, scientific inductive reasoning in science is a restricted form with additional requirements. First, generalization in science can be applied only to the same subject or quantity. Generalization from one type of subject to a different type will be discussed in the following in conceptualization. Second, generalization in science has to have evaluative control mostly from interpolation versus extrapolation. This is consistent with Mill's concomitant method. Qualitative generalization such as based on stereotyping is not scientific inductive reasoning. Therefore, most examples used to develop conventional theories of philosophy of science are *not scientific inductive* reasoning because there is no quantitative control. In science, most inductive reasoning provides quantitative empirical connection or correlation of various factors and variables in order to build our knowledge based on the generalizations of limited observations, e.g., observation-based correction or laws of nature. How to derive empirical correlations is discussed in Section 10.4. It is not based on formal logic and science does not need to explain why, for the moment, such a relationship exists. This is the power of observation!

If the generalization does not result in knowledge-understanding, it is a piece of knowledge-that if the result is robust. As we have learned, knowledge-that does not have much prediction capability and is potentially fallible. Since prediction is from general to specific, it is deductive reasoning in nature that inductive reasoning is unable to deliver. If generalization results in a law of nature, it can be used for prediction as discussed in Section 9.2. However, the prediction is limited within the range of validity of the law and can be fallible in an extrapolation, such as predicting the future or the next observation.

Conceptualization may be considered high-level inductive reasoning that provides connections among many pieces of knowledge-that. For example, Boyle's law, Charles's law, and Avogadro's law each provide the correlation between two parameters from direct observations. Therefore, each is a generalization based on quantitative inductive reasoning. The ideal gas law is a conceptualization combining the three laws. Similarly, Newton's theory of mechanics is a conceptualization combining Galileo's relativity and Earth's gravity measurements, and Kepler's laws of planetary motions. Quantitative conceptualization like this is the backbone of science and is often used as a start point of scientific deductive reasoning.

In general, inductive reasoning without conceptualization does not provide knowledge-understanding.

Without conceptualization and quantitative control, generalization of observations to the same phenomenon is stereotyping, while applying generalization to multiple types of subjects is an analogy in nature. As we discussed in Section 7.2, either stereotyping or analogy can be fallible and is not a form of scientific reasoning. Generalization of these types required conceptualization and is often under careful scrutiny on its applicability with dialectical reasoning which will be discussed in greater detail in Section 9.4.

An interesting example of large-scale correlation analysis is traditional Chinese medicine which was based on large-scale long-term observations of correlations among multiple parameters. The parameters may be grouped into three categories. The personal parameters include the constitution of a person, age, and gender. The constitution may be divided into, say, five values and is one of the controlling parameters. Then there are factors, such as the food, season, and climate of the place, each of which is also divided into, say, five values. From these two groups of parameters, the maintenance of the health conditions can be determined. Then we can deal with the illnesses and medical treatments, which may have three major factors: symptoms, illness, and herbal medicines all as functions of constitution, age, and gender. Medical treatments involve this complicated matrix of correlations purely based on observation. These correlations observed for a few thousand years in one of the most populous ethnic groups in the world are what traditional Chinese medicine is based on.

From a point of view of science, the Chinese medicine *theory* is *one* of the possible conceptualizations and should be separated from the *observational data*. The theory made up a storyline so that doctors could memorize these complicated relationships, thus transforming this information into knowledge-understanding which could contain potentially flaws. This is very typical of inductive reasoning and conceptualization. Its application requires deductive reasoning for a specific patient with a specific set of symptoms. However, there is no specific rigorous deductive-based logic for applications of such a complicated theory, so the personal experience of individual doctors becomes a crucial factor that produces uncertainty in the result of treatment. A failure of a treatment may be blamed on the flaws either in the conceptualization of Chinese medicine theory or in the deductive application due to the doctor's experience, but the correlations of various factors and symptoms roughly hold true because it has had saved uncountable lives over a few thousand years in China and east and southeast Asian countries. Just as any scientific theory could be wrong, the Chinese medicine theory could be wrong or need modifications. The observed correlations can be used as a basis for other new theories. A related example will be further discussed in detail in Section 10.5.

There are many fields of research that are in a similar situation, such as psychology. When no law is rigorously derived via deep-thinking conceptualization, the theories are generalizations based on simple (semi-quantitative) inductive reasoning. The correctness of explanation and prediction depends critically on the robustness of the generalization. If the generalization has a large error or uncertainty or is based on a small number of samples, the prediction with deductive reasoning is not trustworthy. For example, the conclusion that human learning is basically the same as in rats and other animals may be the most "creative" idea one may find. But it is not scientific.

Reasoning Flow: From the left panel of Figure 9.1, it is clear that, in general, with deductive reasoning alone, one is not able to connect two phenomena that are not on the same vertical line of reasoning. If the two are directly connected by single deductive reasoning, it is not a scientific problem but an application problem. Inductive reasoning can go across the lines of deductive reasoning and can connect one law to another, as shown in the horizontal segments of reasoning in the left panel. In science, an observational correlation between two (or more) phenomena/factors is a method commonly used without invoking an existing law and itself can be proven and later become a law.

There are two general possible outcomes from a correlation study for generalization. First, a correlation may be traced to some reasonable expectations of deductive reasoning, i.e., the result of inductive reasoning may be parallel to that of deductive reasoning, in the vertical direction in Figure 9.1. For example, when one measures the volume of a gas with varying pressure at a given temperature, they find that the two are anti-correlated with a nearly constant factor. This is expected according to the ideal gas law. In this parallel case, the result can be used to justify and explain the deductive result. Second, an observed correlation is not expected nor understandable. In this case, the result of inductive reasoning is perpendicular to that of deductive reasoning. In perpendicular situations, the results can often be used as a starting point for deductive reasoning (as discussed at the end of Section 9.2) and/or to simplify mathematics.

The perpendicular situation is more important in science because it contains new information and may lead to the discovery or invention of new knowledge. For example, in the above ideal gas case, one would find that the correlating factor between the volume and pressure is no longer constant when the temperature is extremely low, say close to –273°C. This observation led Van der Waals (1837–1923) to inventing Van der Waals equation of state. This deviation from the ideal gas law can be understood by the finite size of the gas molecules, a new piece of knowledge-understanding.

Potential Problems with Scientific Inductive Reasoning: There are a few problems one needs to be aware of with inductive reasoning in science. First, there are often or always exceptions, e.g., black swans or non-black ravens. In correlation analysis, measurements are scattered around an underlying trend, if there is one, because *multiple effects* may operate. Any conclusion based on inductive reasoning must leave room for them. Therefore, the conclusion would not describe an "all…" or "none…" situation, but rather about the likelihood of something happening before full understanding is developed. It is based on the probability theory of inference which, in philosophy of science, sometimes is referred to incorrectly as Bayesian inference. There are some confusions, misunderstandings, and misinterpretations in philosophy of science concerning the probability theory and correlation analysis that will be discussed in Section 10.4. When the probability is involved, the validity of a result cannot be falsified based on a single counterexample. Therefore, Popper's theory is fundamentally invalid in science because it is based on the absoluteness by required deductive logic of philosophy of science.

Second, two related observations may not be related to scientific causes. Sometimes, associations between these observations may simply be accidental. When a dataset is not very large, an accidental case may carry extra weight to a conclusion that is potentially flawed.

Third, in science, when a correlation is found between two effects, we, in general, do not know the causal relationship between them from the correlation and should allow the possibility for either of the two as the cause of the other, as indicated by the arrowhead at each end of inductive reasoning in Figure 9.1. In some experiments, if time history is included, it may be used to establish the causal relation. However, the two robustly correlated effects could be caused by a common cause, e.g., thunder and lightning.

Fourth, interpretations of correlations are inherently uncertain since a single correlation can draw many different interpretations/conclusions. Some can be inconclusive or totally wrong.[3] Some experimental results

---

[3] During World War II, there were many traffic accidents by US military drivers. To reduce such accidents and unnecessary losses of lives, the US Army requested a leading group in psychology to identify the personality traits of the drivers that caused accidents. The psychologists created a large database of information about the drivers who were directly responsible for accidents and conducted correlation analyses. With the information, they used many indicators to categorize the drivers. What they found was that body tattoos had the highest correlation with drivers that caused an accident. At the time, laser tattoo technologies did not exist; getting a tattoo was a very painful process. Therefore, one possible explanation is that people who got tattoos on their bodies had "a preference for bodily risk" so that they were more likely to cause accidents. This explanation may sound reasonable until you question the logic that leads to the conclusion. It implies that the soldiers sought bodily pain from both a tattoo and a car accident. This does not seem to make sense because it assumes that the drivers planned to get into an accident, seeking the pain associated with that risk, and they would avoid the crash if it would take their lives. Clearly, this, constant evaluation of whether the accident brings pleasure, is not what people are thinking before an accident. Being called 'accidents', people typically believe that there was no problem for them to get through. This is not related to whether the effects of an accident are too painful to take or not or the right amount of pain to be enjoyable. A second possible explanation is based on the driver's overestimation of one's ability and/or underestimation of the danger. This can be related to several psychological characteristics shared by causing accidents and getting tattoos, such as a tendency for showing off. When comparing these two explanations, which one was more reasonable? This example demonstrates that a single experimental result can be interpreted in multiple ways.

are more difficult to interpret. For example, can the high correlation between smoking and lung cancer be interpreted as smoking causing lung cancer? A problem arises due to the fact that not every smoker develops lung cancer while some non-smokers do. Clearly, there needs a better understanding of cancer, such as how lung cancer is caused and develops as well as the roles that smoking can play in the process. As we have discussed, correlation itself cannot be simply used for scientific predictions of the future. These predictions, rather, need to be based on knowledge-understanding.

Fifth, when there are multiple factors of various strengths, some of the factors may be correlated among themselves. For example, education level, intelligence level, and annual income are correlated to some degrees. In a correlation study, it is difficult to separate the effect due to education or intelligence from that due to income. Using correlating variables in studies has been, debatably, abused in some sub-disciplines of social, medical, and behavioral studies. For example, just recall how many times you have been recommended to take or not take a certain vitamin. These confusions were most likely produced by the correlating variables used in a study. Similarly, in some social and political studies, race is often used as a catch-all parameter to characterize an issue. Of course, any result from such a study could show differences among races and the results can easily make news headlines. But in many problems or phenomena, other variables could be dominant factors, such as education, financial status, religion, cultural background, neighborhood norm, party affiliation, etc.; race may not be the determining factor. However, in such a study of a specific issue versus races, the conclusion would be that the issue is determined by race and hence becomes a race issue. In some sense, the current political environment may have been fueled by these abusive/irresponsible studies.

Finally, a potential fallacy of correlation analysis, often discussed by philosophers of science, comes from the so-called reliabilism theories, as described by the "Henry and the Barn facades" example. In this example, Henry drove on a road and saw many barns along each side. He stopped at one, checked it, and found it to be a real barn. He then concluded that this is a barn county, but in fact, there was only one barn that is the one he checked (because it is most accessible), and the others are only barn facades. The reasoning is based on stereotyping of one example, which is the most common fallible non-scientific reasoning. The situations may happen in the common law system where judgments are made based on the previous ruling of a relevant case. The applicability of the previous ruling to a newly brought-up case is not checked. In comparison, in science, research often starts with a case study and is followed by a statistical study. As a scientist, I cannot completely rule out the possibility of relia-bilism in science. Rather, I feel that this is more speculative. If someone shows one example and concludes that everything else is the same, one day there would be a student who will take a look at one or more cases. If the student finds that the example in the case study is fundamentally dif-ferent from the statistical result, the student could overturn the conclusion of the case study and statistical result. Therefore, the conclusion of the flawed study would be a great gift to the student who can be a hero for recognizing that the study *got it wrong*.

9.4 Scientific Dialectical Reasoning

The third type of scientific reasoning is *scientific* dialectical reasoning. We saw similarities between dialectics, from Section 7.2, and the falsification-ism of Popper's theory in Section 4.4. Recall that Popper's theory was strongly disputed by philosophers of science while scientists mostly agreed with it. There is a need to discuss dialectics further in philosophy of science.

Dialectics: In logic, dialectics are known as "minor logic". In contrast, the corresponding "major logic" is in the form of "critique", which requires identifying the weakness or error of a theory. Dialectics is different since it does not need to identify exactly where the problem or error starts and rather simply points out the inconsistencies in a theory; when an inconsist-ency is recognized, it is the job of the proposer of the idea to identify where the problem is. In philosophy, this may not be the right way to argue because one can armchair-speculate anything, but in science, we have a restricted form of dialectics as to how it is conducted. Specific requirements are needed because, otherwise, dialectics can be abused and become armchair speculations. In science, because any idea or result has to be presented in public and has to go through a review or referee process, in both processes, everyone else has the right to challenge it. The proposer of the idea or the author of a publication is obligated to address all reasonable comments or challenges, while the whole process is monitored by the science community. If you were the proposer of an idea, you would have to be prepared for dialectical challenges by others. Therefore, it is better if you could self-criticize your own idea first. To be a good scientist, either the challenger or the proposer has to speak following the established scientific reasoning. Armchair speculations would be dismissed as unprofessional.

The formal form of scientific dialectical reasoning starts with a "reasonable" and "relevant" but alternative assumption to the theory in question. One then uses strict scientific deductive/inductive reasoning, which typically follows the same line of reasoning used in the proposed idea in question, to derive a contradiction or negative conclusion. Note that this type of reasoning is only valid when negative results are derived. It examines an idea from multiple viewpoints/directions based on falsification to eliminate possible ideas. Dialectical reasoning in general is destructive. The detection of an error does not amount to a proof of the antithesis. However, most often, in science, a dialectical debate does not result in one side losing and the other side winning as often in politics. Instead, the new idea can be modified to address the concerns of the challenger. The proposer can defend the idea in order to withstand the falsification. When the assumption made by the challenger is not reasonable or not relevant or the reasoning contains flaws, the proposer does not need to modify the idea.

A new idea that withstands all possible challenges/falsifications, i.e., scientific dialectical reasoning, is tentatively acceptable and has the potential for further development and applications. This should have been the correct form of Popper's falsification theory. The difference from Popper's theory is that the science community DOES gain confidence in the new idea with the scientific dialectical process, although it is still open to further testing or falsification.

This method is most often used at any of the critical connecting points in inductive or deductive reasoning, as indicated by the red dot at the intersection of two lines of reasoning in Figure 9.1. In dialectical reasoning, however, a positive result does not count as justification due to the fallibility of abductive reasoning. In this case, the proof may not be relevant or may be intrinsically circular. It is not uncommon to see in the science community that sometimes inexperienced scientists make assumptions that implicitly include the idea yet to be proven, i.e., circular logic. Other times, the supporting evidence can be an exception, but it is used to support a more general idea. In these cases, dialectical reasoning is a powerful tool to uncover the flaws in reasoning. Scientists will need to learn to apply scientific dialectical reasoning to their own ideas. This is the most important way to address the question of "how to know I am not wrong".

Dialectics won a bad reputation from Hegel and Marx, especially with the Hegelian dialectic triad of thesis, antithesis, and synthesis. In philosophy of science, this triad is often explained with the famous example of light. Newton thought that light was a particle (thesis), while Maxwell showed that light was a wave (antithesis). Then, quantum theory showed that light is both a particle and a wave (synthesis), the so-called wave-particle duality. Marx theorized that the dialectic triad can continue working universally at higher levels, which was then proposed as a universal law of civilization development. The scientific dialectical reasoning, however, does not carry any indication about the next step, either antithesis or synthesis, other than falsifying the thesis. Therefore, the dialectical triad most likely is irrelevant to science.

Scientific dialectical reasoning is investigative and should not be taken as argumentative. In science, there is a particular form of dialectical reasoning, called a "gedankenexperiment", which means "thought experiment" in German. One does not need to carry out the experiment, but rather imagines one and then reasons the consequences, etc. This method is valid only if a negative result is derived concerning a new theory as dialectical reasoning. I need to emphasize that if the result of this reasoning is positive, it can be used to "describe" the idea but cannot be considered proof because, again, it would then be abductive reasoning. To prove a theory according to the gedankenexperiment method, one has to exhaust all possible, but relevant, experiments. This, in theory, cannot be conclusive. When more information is available, more possible gedankenexperiments could be proposed.

Potential Problems of Scientific Dialectical Reasoning: There are a few caveats when using scientific dialectical reasoning. First, it is not conclusive: if done correctly, one can (kind of) know for sure what does *not* work but cannot be sure about what *does* work. Second, in theory, there are an infinite number of possibilities: possibilities of falsification cannot be tested completely, i.e., we cannot come up with an ultimate proof to an idea. Popper's falsification idea is valid to some extent in this sense but is not valid universally or exclusively. Proof can be derived from scientific deductive and inductive reasoning. Third, the problems associated with the steps of both deductive and inductive reasoning (e.g., questionable assumptions, irrelevance, accidental, exception, permeability, leakage, and contamination discussed in the last two sections) in dialectical reasoning may also lead to false falsification. Fourth, because dialectical reasoning does not prove anything but is rather used to disprove something, it cannot be used to determine causal relationships.

However, for scientific problems, we require evaluation to be a part of justifications as well as falsifications. The possibilities of successful falsifications are substantially fewer than those conventional theories of philosophy of science have imagined. Furthermore, many possibilities cannot be tested directly because of a lack of available means. We may also note that sometimes correct or partially correct ideas may be eliminated for wrong reasons because of limited knowledge or flaws in the reasoning process. In these cases, unless the proposer gives upon the idea, they can modify and present it again. However, this cannot be done too often because others would not treat the proposal seriously. When a new idea is conceived, one may need to go back to the line of reasoning to reevaluate the rejected ideas because they may have been rejected for a wrong reason.

Sometimes, wrong conclusions can be drawn from a dialectical debate if a small number of incompetent people are involved. It is quite normal in science for a few scientists to agree on a wrong idea. But from the view of the history of science, one should not worry too much if this happens because as the significance of the matter increases, more and better scientists would critically examine the matter. Other times, debates are futile and do not provide useful conclusions, where no agreement can be reached or is accepted by everyone involved. Some people would accept the position of both sides, both the thesis and antithesis, but, for example, under different conditions or different phases of a process. However, if the information is partial and incomplete, it is quite normal that people are not able to resolve a controversy. These problematic results are unlikely to reach the level of knowledge. Therefore, we do not need to worry about them at the philosophy of science level.

Ideas that seem to be in constant motion and become "unfalsifiable" can be a very difficult aspect in scientific research. In these cases, although the falsifier made much effort, the proposer can always find ways to deflect the falsification. Is this because the idea is pseudoscience according to Popper's definition, or because the idea is correct? Here is a list of issues for the second possibility. First, the relevance between the justification and the underlying truth can be falsifiable although it has not been falsified yet. Second, simple correlations in observations used in falsifications can be accidental. Third, evidence (note, this is not referring to the observation itself, but rather its interpretation) of the falsifications can be flawed. Fourth, the quality of the data may not be able to support the level of specifics that a conclusion demands so the uncertainty in the interpretation is intrinsically large. Fifth, the reasoning of the falsification can be flawed, especially at each critical connecting point. Figuring out the limit of falsifications can often be a more important aspect of scientific research. The method/precision/condition/probability of justification plays a key role in providing any new knowledge as well as in falsification.

Scientific dialectical reasoning is one of the most important and difficult processes in science when scientists challenge a new idea while the proposer defends it in order to "withstand all falsifications". The first issue in the process is the relevance between justification and falsification. The second issue concerns the interpretation of any evidence involved on either side. The quality of the data may be able to support a conclusion to a limited level and the soundness of the assumptions used in the theory constrains its applicability. For experienced scientists, the result of a falsification often limits the range and conditions or improves the soundness of a belief if the belief originally bears truth. The focus is on each critical connecting point in reasoning. Figuring out the limits of the applicability is often THE most important aspect of scientific research (creativity is secondary!).

9.5 Occam's Razor

Occam's razor, as discussed in Section 7.2, is used to break a tie between competing theories, assuming that the simpler theory/idea may have a *higher probability* to occur. Simplicity itself cannot be always used to determine whether a theory is correct.[4] As we noted before, simplicity can be flawed in the cause of the extinction of dinosaurs. The tightness of a theory and evidence can also be crucial. In general, in science, a qualitative theory is often simpler than a quantitative one, but quantitative reasoning may provide much more details, which may attract more falsifications that the theory needs to defend. Therefore, the competing theories have to be similar in nature. When two theories are similar in terms of quantitativeness, we can then compare the line of reasoning. Without mathematical details, quantitative reasoning can be very simple and clean. For example, when studying the motion of the planets, one can base their work on the universal law of gravitation to write a momentum equation for each body in the system. Since the law and the momentum equations are independent of the frames of reference and can be written in a rest frame with either the Sun, the Earth, or a planet of interest, it is possible to study the motion of the planets with either a geocentric, heliocentric, or planet-centric description, as Galileo suspected and as explained in Section 1.4. We like the heliocentric description better (note that this is a "choice") because it is simpler to describe the motion of all planets in the solar system (although the number of assumptions is the same in the three descriptions) according to Occam's razor. This choice does not imply that geocentrism is wrong as it also theoretically describes the motion of the planets correctly. It is a fundamental mistake that many theories of philosophy of science consider geocentrism entirely wrong. Historically, Copernicus made a correct argument for his heliocentric theory. In his defense, he did not argue that geocentrism was wrong but rather argued

---

[4]An example of Occam's razor is the two paradigms in Magnetosphere–Ionosphere coupling, as discussed in Section 6.5 of Chapter 6. In theory, the electric circuit and MHD paradigms are two equivalent system descriptions under most conditions. MHD theory is much more sophisticated than the circuit theory in mathematics, but physically simpler and needs fewer assumptions. But Parker has shown that the electric circuit paradigm is not mathematically trackable, while the MHD theory is.

for the consistency among systematic assumptions needed in heliocentrism when Newton's laws of motion were unknown. This may be considered an argument of Occam's razor.

A few decades ago, when I was a graduate student, there was a popular idea that uses aesthetics, e.g., symmetry, as reasoning to break a tie. At the time, it was very intriguing to me. I did not follow this topic afterward, though I think aesthetics are subjective and person-dependent. To most people, such as Copernicus, simplicity is aesthetic. However, I would not agree to use aesthetics when making a critical judgment of a scientific idea. One way to invoke Occam's razor is the simplicity of line of reasoning, which will be discussed in Section 9.6.

9.6 Scheme of Scientific Reasoning

Two Forms of Reasoning for Scientific Proofs: In science, sometimes, the reasoning of a specific subject can be complicated or obscure. Some philosophers of science take mathematics as a proxy of deductive reasoning, but this may make mistakes. A safe way to identify the nature of reasoning is to see whether the reasoning is reduction going from general to specific or generalization from specific to general. The former is deductive and the latter inductive. For example, to derive a mathematical expression from observational data, e.g., a best fit, is inductive and the application of the mathematical form to a specific case is deductive. Similarly, the conclusion that the lunar phases can be produced when a sphere moves around a point source is drawn inductively, and the conclusion that the moon is a sphere is deductive. A flawless inductive result can be considered a knowledge-that or laws. It is worth mentioning that knowledge-that and laws, in general, cannot be derived by deduction but are often explained or understood with analogies that are based on induction. They can accurately reflect reality, i.e., the truth.

Darwin's evolution theory and Newton's mechanics each were based on a large number of observations of various types, i.e., many pieces of knowledge-that. However, their theories are to reveal the underlying connections among these pieces. To do this, the scientist needs to carry out deep-thinking "conceptualization" which augments or generalizes these results via inductive reasoning — there are uncountable cases that were not observed yet. These conceptualizations lead to milestones in knowledge-understanding development.

The application of a theory to a specific species or a phenomenon is deductive. In the case of black ravens, the hypothesis of all ravens being black is drawn based on inductive reasoning but is not proven. The prediction of the next raven to be black is based on deductive reasoning. However, because the hypothesis is actually not proven and flawed, it is not surprising the prediction is wrong. The fatal mistake in the raven paradox is to use an unproven and flawed "hypothesis" as the start point in deductive reasoning. Even if the hypothesis is proved, because the conceptualization is very shallow and it is not a law nor can provide knowledge-understanding, it does not have (deductive) prediction power. Similar is Russell's chicken–farmer case. Most philosophers of science categorize the two-step logic (generalization and then making a prediction) as inductive, a flaw in conceptualization because prediction is always deductive.

Direction of Science Progress: The objectives of scientific research are to (1) connect pieces of knowledge-that to become knowledge-understanding with conceptualization, (2) reduce components of unsubstantiated belief in knowledge and replace them with inductive and deductive reasoning, (3) reduce and condense inductive components and replace them with more deductive reasoning so that results become more deterministic, (4) present more laws mathematically so that results become more quantitative, and (5) reduce or loosen requirements and conditions with more general (universal) laws.

For example, in step 5, both Kepler's theory and Galileo's theory combined many experiments and observations with mathematically describable laws. Newton's work then combined mathematics from these two theories into more compact forms so that they can be applied to many more problems. In some science disciplines, because the processes may be intrinsically probabilistic, such as in life science and quantum mechanics, the results may be less deterministic and more probabilistic. In these situations, step 3 above needs to be *deterministic probabilities* versus simply random guesses. The deterministic probabilities should be replaced gradually more by deductive reasoning (mathematics) and less by inductive reasoning (observation).

Scheme of Scientific Reasoning: Given that there are three key elements to science, i.e., observation, conceptualization, and evaluation, it is clear that no single scientific reasoning is able to derive a sound scientific result because no reasoning can cover all the needs of the three elements. Each form of reasoning has its weakness or limitations, as discussed in Sections 9.2–9.4. For scientific research, all three ways of reasoning are necessary. Empirical observational results, which are the basis for inductive reasoning and may be combined with deductive reasoning, are often a starting point of research. The key, however, is how to interpret an observation. Deductive reasoning is essential for theoretical research and any interpretation/understanding of an observation. Dialectical reasoning, when combined with deductive reasoning and compared with various observations, is used to examine the input to and outcome from deductive reasoning and at each critical point where the line of reasoning changes. It is essential to answer the question of "how do I know I am not wrong". The peer-review/reply process and seminars and conference presentations involve mostly this reasoning.

The right panel of Figure 9.1 illustrates two possible lines, shown in green and brown, of reasoning/conceptualization/understanding. Inductive reasoning in science may, most often, take the form of a correlation analysis, correlating two variables each on a line of deductive reasoning. When reasoning is conducted this way, one may be able to jump from one line of deductive reasoning to another to provide more possible options for using mathematical tools to solve a problem. Scientifically, an inductive correlation study itself, in general, cannot determine the causal relationship between two correlated quantities, as indicated by the two possible arrowheads in each horizontal segment of reasoning in the figure. However, the causal relationship can be determined according to the overall flow of reasoning in Figure 9.1. For example, there is a segment of inductive reasoning where the green and brown lines flow in the opposite directions. The two lines of reasoning take the causal relation differently. On the other hand, dialectical reasoning is most useful when one considers changing the line of reasoning at each red dot in the figure. At each point, one encounters multiple choices and must decide to go in a particular direction or another. One then has to provide arguments or evidence why not to go in the other directions. As required by dialectical reasoning, one needs to follow each of the potential possibilities with strict deductive and inductive reasoning until a negative result is reached, indicating that the route is a dead-end (according to the theory). If one cannot reach a negative result, one has to seriously consider the possibility as a valid option, even if this is not the proposer's favorite idea.

The reasoning of the green line could be proposed by a scientist who is theoretically oriented. The proposer tries to avoid navigating through a maze of observations. The proposer represented by the brown line, on the other hand, is more observationally inclined. The proposer is familiar with a large range of observations and can easily cite many observed correlations. The mathematics involved in the brown line could be relatively straightforward. The brown line invokes more inductive reasoning and jumps from one point to another to avoid mathematical complexities. The two lines of reasoning start at different points, with the brown line slightly closer to reality than that of the green line. Since their ending points (prediction of each theory) have a similar distance to the observation, more information is needed in order to better differentiate the two. Sometimes, the mathematical treatments used may not be tractable or may encounter some difficulties. Therefore, mathematics alone may not be able to reach a conclusive connection between the two phenomena. As shown in the right panel of Figure 9.1, the first deductive line of reasoning encounters a mathematical difficulty two steps from the starting point. This illustration provides a view of how understanding in science progresses.

According to Occam's razor, the green line of reasoning is considered better because it is shorter and involves fewer changes of reasoning, i.e., a smaller total number of red points along the line of reasoning. If each change of reasoning produces some uncertainty or reduces the probability, the likelihood of success of the green line is greater than the brown line. In actuality, the two approaches may take place in sequences, such as the brown one being proposed first and the green one afterward. Since the green one is logically more straightforward and mathematically cleaner, the science community would gradually shift the understanding/ description from the brown one to the green one. Eventually, the description based on the green line will appear in textbooks years or decades later. This argument may have some flavor of IBE in philosophy of science. However, our criterion and the reasoning in each step are more specific and concrete than what constitutes an IBE. The science progress made from the reasoning of the brown line to the green line is in steps 3 and 4 in the subsection of Direction of Science Progress.

Notes on Scientific Approaches: There are a few commonly used scientific approaches, which some scientists may refer to as scientific methods but are different from those discussed in philosophy of science. These scientific approaches include conducting systematic experiments, isolating variables, analyzing or ranking and prioritizing effects according to order, correlating different independent variables, and searching for alternatives. These are what is outlined by the great scientists in Section 7.1 of Chapter 7. However, using these methods does not guarantee a derivation of a correct idea or an invention of new knowledge.

In science, there is no "so-called" scientific method that is universal and fool-proof, just as most good scientists agree and many philosophers of science eventually conclude. But the sad thing is that the general public has caught up with the wrong idea of scientific method promoted by the early theories of philosophy of science. The flaw of the idea should have been easily proven by a dialectical argument: if there were such a method, all scientists would derive the same (or at least similar) results and ideas. If this were the case, science would not be interesting and not be challenging. The so-called scientific method may be the most influential, but in a negative manner. It is a legacy that the founders of philosophy of science have left to our modern society. Due to the diversity of scientific problems and the limited available observational means and existing knowledge, scientists need to take whatever appropriate approaches necessary to advance human knowledge, as opposed to just an approach supported by the scientific method derived from philosophy of science.

A science project may be motivated by an observation or a collection of observations that may or may not appear to be related. In this case, the objective is to find an explanation or the cause. Since a science problem is not simple, it cannot be proved with stereotyping or analogy. It cannot be proved by abductive reasoning either, backward from the observation to understanding. Instead, one has to conceptualize the explanation and then prove forwardly from the explanation to observation, such as in the case of the extinction of dinosaurs by an asteroid hitting the Earth. A project can also start by application of an understanding to a different observation.

Often, new ideas are not in short supply in science. Instead, one of the most time-consuming efforts of science is to eliminate the "new ideas" that are less likely to succeed. In this process, the potential possibilities to success keep changing. The giants of science are able to speed up this process, mostly with their insight to recognize the potential of each possibility. This insight could significantly depend on the experience or sometimes purely on intuition which could also depend on experience. Most scientific experience can be associated with past successful, as well as unsuccessful, efforts. During these efforts, especially unsuccessful ones, a scientist would have investigated many possibilities before eventually giving upon them. In this process, the potential connections among pieces of knowledge in one's thinking, as discussed in Chapter 3, are important. Experiences gained from failed efforts are not well documented in publications. After conceiving a new and exciting (better if useful) idea, a scientist spends more time in order to get it right, present it soundly, and make sure that "I am not wrong".

The theory described in Figure 9.1 in the green line is based on mathematics to carry most of the deductive reasoning. If there is no mathematics-based law via deep-thinking conceptualization, there are two major potential problems. First, if there is a non-mathematical paradigm, such as the theory of the invisible hand in economics, deductive reasoning is qualitative. This leaves much room for flaws to be introduced, e.g., by contamination and leakage. Second, many "principles" in "newer" science fields are not based on deep-thinking conceptualization. Without such conceptualization, the ideas are stereotyping or analogy in nature, such as the theory of human learning being the same as rats in psychology. They have a lot of chances for flawed conclusions. The simple rationalization of these results cannot deliver what "science" is expected and should not be referred to as a scientific theory or result. The applications of these "principles" to explain a specific phenomenon appears to be from general (animals) to specific (humans), namely deductive reasoning, but in fact, the explanation is more inductive (analogy) in nature — to conclude that all swans are white based on the assumption of all ravens being black. For deductive reasoning, the starting point has to be carefully thought through. Otherwise, it is a garbage-in-garbage-out process. Third, many prediction models are based on the parameterization of historical data. For example, the early weather forecast models were based on a search in the historical records under similar conditions. Similarly, most prediction models of economics or finance are based on historical data. Neural network and artificial intelligence models are also in this category. These forecast models are not based on knowledge-understanding, but knowledge-that. Again, as discussed in Section 9.3, although the predictions appear to be from general to specific because there is no rigorous deductive reasoning for such predictions, they are still inductive. Since the prediction and explanation are not based on knowledge-understanding, they are not based on science.

Questions for Thinking

1. What is the reasoning for Darwin's evolution theory? Is it scientific?
2. What is the reasoning for Newton's theory? Is it scientific?
3. Is the theory of science outlined in this book science, knowledge, or speculation?

Chapter 10

Other Models of Philosophy of Science

Since we are learning about philosophy of science, it would be necessary to, at least, learn about a few theories commonly discussed in philosophy of science nowadays, although I have demonstrated in earlier chapters that these theories were developed under the misguidance of the overall theme of the discipline. Nevertheless, without knowing about them, the education would not be complete. These flawed ideas helped identify the issues discussed in this book. Examining these flawed theories may trigger some enlightening ideas for the readers. I will add my own comments when I see fit during the presentation of these theories. One may challenge these comments.

10.1 Naturalism

Naturalism in *philosophy* believes that the world operates under natural laws and forces. In *philosophy of science*, naturalism does not concern the ontological problem in philosophy and thinks that the best ideas used to understand nature may come from an understanding of the natural world itself. Explanations of the natural world should be natural to our observations and can be tested and verified. It does not need to follow any theory of philosophy or epistemology. In other words, according to naturalism, philosophy of science can use results from science, not just the logic of philosophy of science, to help answer questions in philosophy and in philosophy of science itself. It is worth mentioning that this line of argument is from a philosopher, not from a scientist. However, this idea was immediately attacked by foundationalism, which argues that philosophy of science should be done from "an external and more secure standpoint" to prevent the possibility of producing circular logic in studies. According to foundationalism, the theory of philosophy of science should be transcendental to all sciences; this is what the traditional philosophy of science has done. The result is that it becomes irrelevant to science while many philosophical questions from science are not even discussed, as discussed in Chapters 8 and 9.

An influential work is the 1969 book entitled *Epistemology Naturalized* by W. V. O. Quine. Quine (1969) challenged the usefulness of the traditional philosophy of science based on the failures in finding an answer to the criticisms of induction and the inability to develop a universally applicable scientific method. Quine notably challenged philosophy of science for its failure to derive a purely deductive logic that can replace the inductive methods widely used in science. If there is no direct unambiguous relationship between scientific knowledge and its logical structure, all ideas of linguistics and logic analyses done by positivism are not useful or relevant. Quine's theory claims that any attempt to give a general philosophical foundation for science is always doomed to fail because science does not need a philosophical foundation in any case, unlike the claims traditional philosophy of science took as their pride. Quine's work is often considered the final blow to the traditional philosophy of science discussed in Chapter 4. In this regard, naturalism derives a similar conclusion as derived in Chapters 7–9 of this book. I agree with his diagnosis but disagree with his prescription below.

As an argument against the objection of circularity of the theory of naturalism that uses science to explain what science is, naturalism suggests that science is determined psychologically. Furthermore, in this case, epistemology is not judged by truth but by our psychological judgments of truths based on a natural existence that can be observed and tested. Psychology should eventually be able to give a purely scientific description of how beliefs are formed and how they change. However, this argument invited more objections because it appears to be in the reverse direction from reductionism to be discussed in Section 10.7. Many philosophers of science argue that science starts with a physical under-standing of the natural world, and then moves on to chemistry to biology, and then to psychology. According to the theory presented in this book, there is confusion in either approach, physics to psychology or psychol-ogy to physics: Psychology concerns primarily the problem of conduct and not the problem of knowledge. We still do not know how to scientifi-cally model the problem of conduct.

Within naturalism, there are two major groups: normative naturalism versus instrumental rationalism. The former believes that philosophy of science should involve a value judgment, a judgment of what something "should be" versus what it "is". However, as we discussed before, how to decide what "should be" is highly debatable because there is no com-monly agreed "should be". As a scientist, I completely reject any philo-sophical idea or notion of normativity when discussing the development of science because science is searching for truth. Truth is objective and does not care about whether philosophers think it "should be" or not. On the other hand, the latter thinks that value judgments are not needed and how to achieve a goal is the only thing that matters. What that goal is or whether the goal is good is not a problem for philosophy of science. This would be accused as goal-oriented by some.

Instrumentalism sometimes is referred to as against realism which will be discussed in the next section. To the instrumental rationalists, the problem is whether observations can be trusted. When science becomes complicated, they argue, many scientific understandings are not intuitive. Therefore, scientific theories may be only useful instruments, heuristic devices, and tools to organize our experience but should not be taken liter-ally as being true or false. Instrumental rationalism is ready to invoke a theory if it is useful or reject it if it does not work. In this case, observa-tions are often not a direct reflection of the natural world but are based on some theories (called theory-ladenness). Therefore, an observational result can be interpreted differently. Although this conclusion is the same as we discussed in Chapters 7 to 9 in some respects, Quine concluded dif-ferently. According to Quine's theory, we cannot test a single hypothesis or sentence in isolation. Any test is a test of a network of hypotheses. This is famously called "holism" that has been discussed in Section 4.5. Naturalism concludes that traditional empiricist views of the roles of observation in science are flawed and simple falsificationism does not work for scientific judgments. Although the conclusion is again the same as ours, the reason for the conclusion by the holists is clearly flawed. It is true that to a philosopher, any test of a scientific idea involves a network of hypotheses that are out of the expertise of the philosopher. However, the verification of the applicability and accuracy of these hypotheses, as well as the resulting uncertainty, is not the task for the philosopher. As described in Section 8.3, the review process of the technicalities is carried out by experts who would be eager to identify the flaws in the test. Therefore, a theory of philosophy of science cannot be based on generic speculations of perceived potential flaws of the technical details that philosophers do not understand.

The original introduction of the term "theory-ladenness", by Kuhn, seemed to be based on his observation of the fact that new observations can be interpreted differently by various theories. Each interpretation sounds plausible and, however, contains theoretical or technical terminologies that philosophers cannot understand. From the examples given by later philosophers of science and sociologists of science, "theory-ladenness" appears to imply that science cannot prove things concretely because of the complexity of the technical and theoretical details involved, such as holism is based on. Naturalism did identify correctly the problem of multiple interpretations of a single observation. However, it tries to resolve the problem within the frame of the traditional theory by questioning the trustworthiness of observation and empiricism.

We discussed this multiple interpretation problem and explained how to make the judgment in science in Section 9.3. In science, each potential interpretation is evaluated by leading experts who are, in principle, neutral and capable of understanding the technical details involved. In principle, if one is not sure about the instrument, the observational data cannot be published, and if the data have a large uncertainty, it will be discussed in the publication. The inability of science to single out a winning interpretation is due to uncertainty and incomplete or insufficient information at a time, not because scientists do not know where the problem is. Therefore, the term "theory-ladenness", as used by philosophers of science, is a misunderstanding and misinterpretation of the cause of the difference between "fact" and "evidence". New successful scientific theories, new technologies, and new inventions prove that the system of science works and is trustworthy. On the other hand, the theory of theory-ladenness underestimates the urge, courage, and ability of scientists to search for truth. With quantitative control in science, the chance for true "theory-ladenness" is very small. Here, I want to emphasize, again, that scientists are aiming at either inventing something new or identifying something flawed. Identifying someone's mistake is much easier than inventing new knowledge. Among experts, someone's misinterpretation due to "theory-ladenness" would be an opportunity for everyone else. Naturalism's suspicion of scientific observations was misguided because the real issue is that "evidence" is an interpretation of an observation.

One of the key questions that naturalism is set to answer is the following epistemological question: What is the relationship between common-sense knowledge and the scientific descriptions of our real contact with the world? This is the relationship between knowledge and science that we discussed in Chapter 8. Science is knowledge in development and has more chances to be wrong than knowledge does; however, knowledge may be older and need an update when scientific results are mature.

Another issue that is under active research by naturalists questions the boundary between two science branches. They envision that they patrol the relationships between adjacent sciences and occasionally climb into a helicopter to get a synoptic view of how the pieces fit together. This self-assignment may be a task that philosophers are unable to carry out successfully. Throughout this book, I have demonstrated a few examples where philosophers of science confused themselves on scientific issues, such as calling Copernicus's theory a scientific revolution and claiming Newton's theory has fallen. In science, the boundaries between disciplines are interdisciplinary studies that are among the most active research projects at the time, such as the problem discussed in Section 6.5. The boundaries are intertwined with fusion processes from both sides. The problem may be too complicated for most philosophers to understand unless they are technically leading experts on the scientific issues. Before philosophers of science reach more commonly agreed upon theories of philosophy of science, it is difficult for philosophers to advise scientists on how to conduct science and what it is. This is similar to a situation where there are no commonly agreed upon traffic laws shared among highway patrol officers. The highway patrol officers would be arguing in the helicopter among themselves when an event is happening on the highway. The intention is good, though. The boundary problem is mostly monitored by (philosophy-loving) scientific leaders.

Naturalism has cast doubt on philosophy of science because scientific methods could be used in philosophy of science; consequently, philosophy can become a branch of science. But there is no scientific method in science for philosophers to use as discussed in this book. On the philosophical level, science has been successful in addressing the problem of *knowledge* but not necessarily on the problem of *conduct*. There may be special issues concerning the problem of conduct that cannot be treated using the same concept as science.

10.2 Scientific Realism

Scientific realism does not primarily concern how to derive something true in science or whether something is true or not. Instead, it mostly concerns the relationship between science and reality, i.e., whether science reflects reality. Note that this concerns *science*, not *knowledge* itself. In my view, scientific realism is more of a philosophical belief or a general epistemological statement. For example, according to common-sense realism, we all live in a common reality that exists independent of how we think about it because we may think about it differently. Scientific realism is directly against instrumental rationalism, believing that the world as described by science is the real world, not only instrumental. The conflict concerns some of the modern scientific ideas, such as subatomic particles, that cannot be directly observed even with modern technologies. Scientific realism bases their argument on the successes of science which imply that scientific theories are related to truth/reality. For example, the debate over the existence of electrons and atoms seems to have proved scientific realism to be right. For untested theories, scientific realists are ready to accept them as reality before they are falsified. For example, an electron is a ball in most scientific models. One may question whether it is exactly a ball or not. Realists think it is. We recall that instrumentalism thinks scientific theories are only useful instruments, heuristic devices, and tools to organize our experience and not necessarily reality. But the debate between scientific realism and instrumentalism misses the point, according to my view: Science is the process to invent new knowledge; knowledge is subjective and is an image of reality. How well an image reflects reality depends critically on the quality of the camera and signal processing. In that sense, scientific realism has more chances to encounter problems. Larry Laudan (Gutting, 1980) compiled more than 30 examples, such as the existence of phlogiston or ether, to which the application of realism was wrong.

In theory, scientific realism writes a blank check to science for whatever scientists do, which is nice for the enterprise of science. However, it does not help scientists in their effort to explore the problems with existing theories or to push human knowledge forward, especially when some appealing theories are wrong or do not correctly describe reality. For example, the caloric theory considered heat a kind of substance. The heat substance resided in a body and could flow from a hotter one to a colder one, like a fluid. Nicolas Leonard Sadi Carnot developed the principle of the Carnot cycle which was used as the basis for designs of auto engines according to the caloric theory of heat. The heat substance should be a kind of matter that one should be able to catch. For example, the theory thought that there should be a lot of heat substances inside matter like metal. When drilling metal, the heat substance would come out. This would prove the existence of heat as a substance and explain why the drilling bits and shavings are very hot. A problem arises when the drilling bit is dull; the metal and drilling bit can become extremely hot but do not produce more shavings. Eventually, it was James P. Joule's experiment that showed heat as a kind of energy, not a substance.

More problems for scientific realism arise from scientific research because often science does not describe the reality that people can sense and more often describes speculative ideas. Is a black hole or a quark a reality? We do not have a clear sense or common-sense knowledge of each.

There are debates within scientific realism. One side is called optimistic scientific realism, which thinks that we can be confident in science for continued success in uncovering the basic structure of the world and in understanding how it works. The other side, which is called pessimistic scientific realism, is more cautious or slightly skeptical. According to the confidence we have in basic physics, the assumption is that low-level structural features of the world have been captured reliably and mostly deductively by our models and equations. But we have not yet found powerful mathematical formalisms in other fields such as biology, psychology, or medical science. Because fundamental ideas have changed so often in science — especially in physics — we should expect some of our current views to be wrong! What a pessimistic conclusion! Nevertheless, pessimistic scientific realism thinks that there are issues with the level of optimism held for well-established theories.

Scientific realism confuses science with knowledge. Science, the stage of knowledge in development, only partially reflects truth and reality at a given time with incomplete information. Scientific realism may be useful to scientists in investigations that are remote from reality, such as studies of quarks and bosons because it gives faith in these studies. In my view, it is likely that the eventual knowledge from these scientific studies may be substantially modified versions of our current theories, as expected by pessimistic scientific realism. Intermediate versions of a theory could be wrong or partially wrong or incomplete, e.g., the idea of the caloric theory or ether, either of which is unobservable but in fact does not exist according to our current understanding. A difficulty scientists face is that they do not know whether the version they are working on is the final version. Although we believe that science has to and will eventually reflect reality, it does not guarantee that a specific science idea reflects reality. In fact, the chance for a theory to be wrong is much greater than to be right, as we discussed earlier.

Only Popper and some naturalists agree with scientific realism; Kuhn and many sociologists do not. Scientists have to constantly make decisions among multiple scientific ideas. We cannot assume that all these ideas reflect reality. A blind belief that science reflects reality does not help scientists in doing science. In science, observation is the truth bearer in a situation, but the multiple potential interpretations of it have to be mostly wrong or need substantial modifications.

10.3 Sociology of Science

Because science is a social enterprise or a social process and sociology is the general study of human social *structures*, sociologists also invented some ideas about science from social perspectives. We should note that sociology may *not* be logic-based but based more on interpretations of observation. An immediate problem here is whether the interpretation of observation in sociology is correct or reasonable. Some sociologists think that sociology may dominate philosophy in real life. Others who write about sociology of science may also consider themselves philosophers. Therefore, their ideas are discussed in this chapter.

General Ideas: Sociology of science believes that gender, money, prestige, politics, and intelligence should not be a factor in science, a belief that I also support. A belief that I have real trouble with, however, is that it thinks science does not serve the whole of mankind but primarily serves certain people, e.g., rich or poor people because a new result from science benefits one group more than others. I believe that science does not belong to a class or a nation. It is for the whole of human society. Philosophy of science, in contrast, focuses on how to get science correctly. As to which sciences the government should support or how scientific results should be used, these questions are unrelated to and distract scientists from their major concerns. However, here I will use some space to describe some of their ideas to broaden the scope of perspectives for educational purposes. These ideas demonstrate that there are people who think about science differently, different from everything that has been previously discussed in this book.

Sociology of science thinks that philosophy of science is misguided by asking big (wrong) questions that it cannot answer — the same conclusion I drew in this book but for different reasons. Sociology of science, rather, asks a different set of questions. Older sociology of science studied the structure of science and its historical developments to find the *norms* of science and the *basic values* that govern scientific communities. It identified a few principles that are supposed to distinguish science and scientists from other social processes and groups of people. The first is universalism, according to which personal attributes and social backgrounds are irrelevant to the scientific value of a person's idea. The second is communism, which regards the common ownership of scientific ideas and their results. The third is disinterestedness, according to which scientists are not working for personal gain. The fourth is organized skepticism, which involves community-wide patterns of challenging and testing ideas, versus simply trusting them. Ideally, I support these principles although I do not think there is systematic scientific observation to back them up. They are more idealized or romanticized beliefs. When discussing specifics, these principles may soon be in conflict with each other or conflict with reality.

The old theories of sociology of science discussed science's reward system (Merton, 1973). According to the theory, scientists are not driven by curiosity but by recognition and reward, contrary to the claim of "disinterestedness". The evidence is that there were a few examples of ugly disputes over precedence and plagiarism about inventions. This is a very good example of *uncertainty in the interpretation* of a fact based on inductive reasoning: Inductive reasoning cannot be used for causal relations. Speculating on the motivation based on result is abductive reasoning which is not valid to prove an idea. It is perfectly reasonable for someone who is driven by curiosity to fight for the credit one deserves after a result is published. Therefore, fighting for the right of the inventor says nothing about the motivation of the invention. The theory of sociology of science which argued for a system of recognition and punishment in science is reasonable but is not derived from reasoning. The theory promotes recognition to the first who publishes or patents a new idea and suggests the best scientists serve on national committees and/or review panels, while those who conduct fraud, plagiarism, libel, or slander are punished. This sounds alright although inconsistent with the four principles and reverses the motivation and causal relation — scientists, driven by curiosity, expect deserved recognition and awards after first inventing a great idea.

When implemented, however, the reward system in some fields has evolved into a so-called "publish or perish" mentality, i.e., one has to keep publishing ideas or lose their job. The publish-or-perish policy adds tremendous pressure on young scientists. The problem is that there is no quality control in this unwritten policy. As a result, a lot of junk has been produced in the process. Some individuals take aggressive approaches against the quality control of a journal by using social media, such as twitter. This action is a new phenomenon that introduces interference to science with outside forces via social media. Currently, there is no mechanism to address this problem. Because the journal publishers have different agendas other than science, when such an event occurs, the publisher tends to deal with negative tweets as a public relations crisis management issue. They would be willing to sacrifice anything, including the scientific integrity of the journal, to calm down the situation.

Recognizing the problems with the old theory, new theories of sociology of sciences use sociological methods to explain why scientists believe what they believe, why they behave as they behave, and how scientific thinking and practices change over time. All these sound interesting questions if studying properly. The new theories employ the principle of symmetry, according to which, the *explanation resources* for beliefs should be the same on two sides of a debate. Fair enough. Note that this principle is based on *beliefs* and not on facts or observations.

Scientists, according to the new theories, are not a special breed of pure and disinterested thinkers and their behaviors are more tribal. Each science community has a local (as in a discipline or subdiscipline) norm that regulates the beliefs and disagreements. Scientific ideas can reflect the interests of a social and political group and benefit that group. Each scientist knows only relevant local norms, not outside norms, but these local norms can only be justified via outside and non-scientific norms. The idea contains something interesting that may be partially reasonable if the terms "social and politic" are not included. At least at the disciplinary level, scientists are not grouped according to social and political beliefs, my observations indicate. Scientists are grouped according to paradigms. What the theory underestimates is the drive and courage of individual scientists to invent new knowledge; a new paradigm may grow out from an old one. A tribal organization or rule has to be able to handle a breakup of itself if the tribal theory works. The science community has very limited influence on individual scientists' behaviors because individual scientists are *not* employed by the science community.

Construction of Scientific Facts: One of the ideas sociologists of science proposed is that scientific facts are "produced" in the laboratory, i.e., not in nature, as described by the book entitled *Laboratory Life: The Construction of Scientific Facts* by Latour and Woolgar (1979). This book, which is better called a novel, was based on visits to a molecular biology lab, work from which resulted in a Nobel Prize. Its observation may be better referred to as a "story". The original idea, or rather "the plot of the story", may sound interesting: to study the science process as an anthropologist studies a tribe that is completely strange to the anthropologist. In science, we often are in a similar situation when we encounter a problem that we initially have no knowledge or clue about. However, one may question whether the *analogy* between science and a tribe is valid, as we discussed in Section 7.2. Note that science is a process whereas a tribe is an organization. In a tribe, the chief has much power over the tribal members. But in science, every scientist is an independent thinker and obeys no authority as we discussed in Chapters 7–9. Nevertheless, from this story, one can imagine and learn about how sociologists would view scientific research activities and, especially, how they draw their conclusions.

Interestingly, their first impression in the lab was that "the heated debates in front of the blackboard are part of some gambling contest". To an ignorant observer, scientific discussions may indeed look like a gambling contest since debates are often intense when each side defends its idea and challenges the opposite ideas. However, in a debate, each side has to follow the scientific reasoning outlined in Chapter 9; therefore, it is not a simple gambling contest. Because knowledge is limited and information is only partial, the idea on each side may have some uncertainty that may result in an impression that scientists are gambling on something. However, as I described in the book, debates of this kind are very productive in science because they crystallize the core differences of the ideas. Scientists can each go back looking for evidence — facts with interpretation — in order to resolve the problem. Therefore, the observation and description made by the sociologists is reasonably accurate. However, their interpretation is deeply questionable which we will discuss later in this section. Let us first finish the story.

As the sociologists observed more, they found that most of the discussion was centered on publications; most of the scientific activities can be summarized as producing papers for scientific journals. Therefore, without an appreciation of their contents, papers are a product of science. Naïve, but this general summary may be a vivid and faithful description of some disciplines of science. Don't sociologists also produce papers and books? As I discussed in Chapter 3, the material form of science and knowledge is eventually publications that can be passed along to future generations. This is not something special to distinguish science from other fields of research.

In their observation, one thing struck the sociologists deeply. The scientists seemed to view the "desire for credit" as secondary, e.g., scientists may contribute their ideas toward the solution of a scientific problem without expecting to be the coauthor of the final product, the paper. This is generally true in science. If this was a great surprise to the observers, the only conclusion is that, in the discipline or the community of the novelists (sociologists), people must have jealously guarded their ideas unlike in the science community where ideas are discussed more freely. At least, in the science community where I conduct research, most scientists share a common set of scientific ethics, and a code of conduct — coauthorship may be offered if the person's contribution is substantial to the final result. But coauthorship is rarely a precondition for a scientific debate, even if intense and lengthy. If a bad incident occurs, e.g., when a debater takes the idea from the proposer to publish a paper without acknowledgment of the original proposer, people simply avoid working with this bad player. Therefore, I take this observation as a compliment to scientists.

Instead, as sociologists observed more, they found that scientists view scientific "credibility" (not "credit") to be more important. Scientists often evaluate who said it when evaluating what was said. What surprised the authors of this novel is that "For a working scientist, the most vital question is *not* 'Did I repay my debt in the form of recognition because of the good paper he wrote?' but 'Is he reliable enough to be believed? Can I trust him/his claim? Is he going to provide me with hard facts?'" I have to say that the authors really got this correct, better than all conventional theories of philosophy of science. Yes, scientists care more about the trustworthiness of a work or author. But I am stunned by the implication of this observation. The norm in the authors' field appears to first consider whether one should "cite" a *good* work before caring about whether the work is *correct*. However, their observation and theory only correctly catch part of what the scientist is thinking. In scientific discussions, although we mention names, we really are referring to a specific work/model/idea of the person. When the issue becomes uncertain, the author's previous works (credibility) may lend support to a judgment. A good scientist *earns* a reputation or "credibility" through careful and insightful works. A debate is often around only a handful of ideas or papers. Better works or ideas are treated more carefully and more seriously. If the science problem remains outstanding, which is the reason for a debate, it is because none of the ideas satisfactorily addresses the problem. Scientists work under conditions of limited knowledge and incomplete information while there are uncountable ideas that have been proposed and investigated on the problem. Under this condition, the credibility of the proposer of an idea, of course, carries weight. A scientist has to examine the works more carefully before modifying or rejecting an idea from a trustworthy scientist.

However, the credibility of a scientist is earned by the work one has done and does not necessarily depend on the school where they received their Ph.D. degree or their job title or the name of their adviser, not as the sociologists indicated in their novel. I have encountered a few low-quality scientists from prestigious institutions in my career. Their home institution does not add more credibility. This was why I stated earlier that a good scientist cannot make major mistakes throughout their whole career. In this respect, the theory that science should be a free market, as proposed by Feyerabend, would not work unless the concept of brand-naming, i.e., products of high quality, is also incorporated. Scientists simply have no time to read junk papers. I'd guess that philosophers would ask who should decide the brand-name of a product. The answer is as follows: by each individual scientist. Since there are only a few works that would be directly related to a study, the scientists would carefully study and make judgments on each of them. People who repeatedly make wrong judgments on the "brand-name" would not be productive in science. Of course, different scientists draw different evaluations, which is one of the reasons for the debate. With these interesting and reasonably correct observations, the authors of the novel, however, drew a completely surprising picture about science.

According to the authors, the laboratory is a kind of machine with inputs and outputs, in a process aimed at *making scientific claims*. It should have been stated as "inventing new science and knowledge" if the authors were not biased nor had a hidden agenda that will be revealed next. They think that scientists built structures of support around their machine so that outside people would not be able to fully understand what is going on. Because of these shielding structures, a process is hidden to the outside that some manmade things are turned into facts of nature. To turn something (that is not a fact but is manmade) into a fact, or reality, according to the authors, is to make it look not like a human product but be given directly by nature. (There is a potential logical jump here. We know that all knowledge and scientific ideas are subjective and are, strictly speaking, manmade in some senses. But inventing an idea/knowledge is not making a reality.) According to the sociologists/novelists, experiments are expensive public relations (PR) exercises. But the opposite idea can also run a PR campaign, can't it? The sociologists have an answer to this question: When one explains why one side of a controversial scientific debate is successful and the other side failed, the winner does not need to give explanations in terms of nature. Instead, the facts made by the winner are immune to challenge, i.e., by having better "PR". This may be their version of what we referred to as "withstanding all falsifications". According to the sociologists' theory, science is controlled entirely by human collective choice and by social interests — the last item is clearly flawed because many scientific inventions resulted from individual activities and not social interests involved. According to the authors, what makes science run is negotiations, conflict resolutions, hierarchies, and power inequalities; the facts are manmade and there is nothing related to searching for truth.

This theory (its reasoning and conclusion) serves as an excellent example to demonstrate the importance of the two central questions that philosophy of science should have addressed (How to get it right? And how to know I am not wrong?). How does one get it right according to the novelists? Observation. In principle, it is a valid way, but there is a problem. In their book, observation was conducted as an anthropologist observed a tribe. This is the same methodology, as Feynman described, of an ornithologist observing birds. Anything wrong with it? In both cases, the implicit assumption is the superiority of the observer's intelligence over the subject of study. Namely, the anthropologists and ornithologists can understand the tribal members and birds but not the other way around. Can this make a difference? Yes. If the novelists wrote an autobiography of Einstein or Newton, would they assume Einstein or Newton was

272 Philosophy of Science

gambling with Newton or Aristotle, respectively? The authors would not speculate negatively on the correctness of the things that Einstein or Newton did even if the authors could not understand or appreciate them. The two sociologists who authored this book clearly have a bias against science and scientists.

Now, the authors also drew a conclusion of a conspiracy about science that all scientists are deceivers and are not driven by their curiosity about the natural world. They argue that scientists are not searching for truth but fake something followed by a PR campaign. We all know that when one builds a big lie from a false fundamental assumption, it is difficult to break. However, for the theory to work, these deceivers are united and have carefully organized the scientific community in which all deceivers decide and agree on the same thing each time, i.e., the same lie, even if a fraction of the deceivers have to acknowledge being the loser of a debate. Did the sociologists ask the question of whether they could be wrong? Clearly not. This conspiracy theory may sound reasonable, but there is a problem. In this scenario, all deceivers seem united and ready to sacrifice their work and reputation as being the loser for the greater cause of the whole community of receivers. They seem more like a martyr than a deceiver! Don't these deceivers deceive each other within their organization? Wouldn't the loser in a deceiving game choose to go public and reveal the dirty tricks of the other side? If they do, they could become a hero to the general public and not a loser. One may note that this line of argument is what I described in an example of dialectical reasoning: assuming the assumption as being true (all scientists are deceivers) and finding the negative conclusion (the loser of a debate has to be a martyr).

I conclude that this book is for entertainment, not academia. A novel with a story line like *Laboratory Life*, however, may be a difficult sale to teenagers because of the self-contradiction discussed above. Studies of this kind could seriously damage the reputation and seriousness of a field of research, such as sociology of science. With so many serious flaws in an influential work, I would be cautious with regard to any conclusions drawn in sociological studies from now on. I recall the reason for caring about one's credibility in science. If sociology wants to be called a discipline of science, it still has a long way to go. It has to learn how scientists think and it is not in a superior position to judge science, if not for entertainment purposes.

10.4 Probability and Bayesianism

According to Popper's falsification theory, a single counterexample can falsify an idea when the prediction of an event is a yes-or-no question. The introduction of probabilities to the predictability of an idea was great progress to philosophy of science. When a prediction is made in terms of probability, it becomes much more defensible while sounding more scientific because of the involvement of math. The probability theory has become popular in philosophy of science, especially because quantum mechanics is based on solving a function of probability. There is no doubt that the indeterministic nature of quantum mechanics has raised many philosophical issues on the nature of science.[1] However, many conventional theories of philosophy of science have confused and misunderstood the philosophical meaning of probability. There are three different applications of the concept of probability in science.

Probability Theory in Science: The first scientific application of probability is essential to inductive reasoning. In science, most phenomena are not determined by a single effect. When *multiple* effects are involved in a phenomenon, depending on the relative magnitude of each effect, a given single effect may not be able to explain *quantitatively* the repeated observations of a phenomenon because every case can be different to a certain degree due to other effects. This raises the probability issue and the prediction based on induction can be presented in a range. The generalization of the limited number of observational results then faces two

---

[1] Some philo-scientia philosophers exaggerated the uncertainty principle in quantum mechanics and argued that because the physical world cannot be determined precisely, humans should not be morally responsible for any wrongdoing as uncertainty is out of human control. Philosophically, the uncertainty principle in quantum mechanics can result from an additional effect that has not been recognized. There is a possibility that an additional factor at the quantum level has not been identified as indicated by Einstein's famous quote that "God does not play dice with universe".

questions. (1) How representative are the finite observations in the overall situation? (2) When predicting the next observation, how different is the observation from the existing ones and how much can the next observation change the overall situation? These two questions are related to most data analyses. Note that we make no assumption here about the correct answer to either question. Hints to the answer will be provided by the data analysis.

Probability theory can answer these two questions. This is one of the most important uses of statistics or probability theory in science. It substantially strengthens the quantitative inductive reasoning used in science. Furthermore, in science, the probability is seldom produced like the "chance" of the head in a coin toss, because it may be caused by multiple potential effects that occur with different phases and strengths which, in plain language, means the relative strengths of the multiple effects at the time of measurement. The fluctuations associated with multiple effects involved in the observational data are expected to be scattered away from the overall trend of the relationship described by the dominant effect. Deviation or scattering of measured points is often taken as caused by "noise" which can be assumed to be either added to or subtracted from the theoretical prediction, although some effects may not be "noise". Philosophically, the scattering in the data of this type cannot be used to falsify inductive reasoning — Hume's theory attributes the results associated with multiple effects to inductive reasoning or non-uniformity of nature, a fundamental flaw.

The second application of probability in science is the statistical generalization to predict the occurrence of an event in terms of probability, i.e., the chance for something to happen, similar to Mill's method of agreement. The difference from the first application is that the correlating factor is qualitative so Mill's concomitant analysis (of magnitude as a function of a factor) cannot be made. Such a study shows only the probability of the occurrence (yes or no in percentage), but does not provide information about the cause or understanding, such as in the observation of the behaviors in the elevator. This is a quantitative version of the stereotyping, inductive reasoning in nature as discussed in Sections 9.3 and 9.6. This application may be important in the discovery phase of knowledge-that.

The third application may be considered a special case of the second application, but its objective is to evaluate the trustworthiness of an idea or proposition. Thomas Bayes (1701–1761) noted that when more experiments are continuously made, the likelihood of the proposition to be true can be estimated based on the "conditional probability" in mathematics which is the probability when a specific condition is satisfied. For example, if a phenomenon occurs under a necessary condition, the probability of the necessary condition to occur can be included to increase successful predictions of the phenomenon. Some philosophers of science borrowed the idea and developed a theory of philosophy of science of using the conditional probability theory to determine the likelihood of a scientific idea. According to the theory, an idea that withstands more falsifications would be more likely to be successful, a conclusion that argues against Popper's claim that successful tests cannot increase the chance for a proposition to be true. The theory is referred to as Bayesianism. Note that Bayesianism does not simply refer to the generic use of probability in science as discussed in the first two applications. Once, one of my colleagues mistakenly referred to the data scattering discussed in the first application above as due to Bayesianism, a confusion.

Potential Problem with Bayesianism: According to Bayesianism, in the process of repeated tests of a proposition, the conditions when the proposition is true can be used to predict the success of the proposition. In this case, if the prediction is correct under a condition, the conditional probability increases. The success rate of the predictions with this condition increases. In a real problem, the conditional probabilities cannot be theoretically determined and are "assigned" by the user, which eventually is translated from the degree of belief of the user (Rosenberg). Therefore, conditional probability is not an objective measure in actuality. It is clear that this theory already deviates from the original idea of seeking truth with science. As we have shown in Chapter 9, to a scientist, accepting or rejecting a theory is based on reasoning and available information and cannot be based on conditional probability. After all, science is not a gambling game.

The third application, Bayesianism, is more often invoked in conventional theories of philosophy of science but seldom used in science.

The second application, quantitative stereotyping, is often used in "newer" fields of qualitative research in the data collection phase. Only the first application, analysis of scattering data, is widely used in science. This creates an extremely confusing situation for philosophers of science as they may consider the three the same. Because inductive reasoning is essential in science, I will use the rest of this section to discuss the idea of statistics and probability used in science, the first application, and how it relates to inductive reasoning although the discussion appears beyond philosophy of science. Nevertheless, this is where many conventional theories of philosophy of science started their flawed argument.

Correlation Analysis: The assumption in data analysis is that *multiple* factors/effects are contributing to an observation. In an analysis, factor X is used as a proxy of the *dominant* effect governing the phenomenon Y. If there were no other effects, there should be a relationship of $Y(X)$ which can be presented as a curve on a chart. There should be no scattering in the data and all observed parameters $Y_i(X_i) = Y(X_i)$ should fall onto the curve $Y(X)$, where subscribe i denotes the ith measurement. However, because of other effects, the observed point $Y_i(X_i)$ would be somewhat different from prediction $Y(X_i)$. The actually observed value of Y_i could be either greater or smaller than $Y(X_i)$, i.e., above or below the curve. If other effects are independent of X, the chance for the observed value Y_i to be greater or smaller than $Y(X_i)$ is the same. Therefore, an observation point has equal possibility to be on either side of the underlying trend $Y(X)$.[2]

---

[2] If the data scatterings are random, there should be more data points closer to the underlying trend and few data points farther away. The positive and negative scatterings, $Y_i - Y(X_i)$, are expected to cancel each other out so that the overall trend of the fundamental, or lower-order, effects can be preserved. The totally random fluctuations can be described by a "normal distribution function", or a bell curve, in the probability theory, which can be characterized by three parameters: the total number of data points, the overall scattering away from an underlying trend, and the average value. A greater number of data points increase the robustness, or trustworthiness, of a statistical result and greater overall scattering indicates that other effects are important. The chance of a data point falling within a certain distance from the trend can be determined using probability theory to make predictions according to this probability.

As we discussed in Section 9.3, since we, in general, do not know the exact causal relation between X and Y, one can swap the two parameters resulting in a different relationship, $X(Y)$.

The amount of total scattering is described by the correlation coefficient (cc.) between X and Y, with cc. = 1.0 indicating that there is no scattering and cc. = 0 being indicative of uniform scattering that is distributed everywhere with no visible underlying pattern on the chart. The correlation coefficient is the probability of the correlation of Y to X compared with other effects. Now, most calculators, computers, and data analysis software packages include functions for statistical and probability analysis. It provides, numerically, the correlation coefficient between X and Y and the parameters for the underlying trend. In general, when cc. is above 0.5, X and Y are considered correlated. In such an analysis, one assumes that all other factors have to be random, i.e., not correlated to X so that these factors cancel each other when the sample size is large enough. If cc. is smaller than 0.5, on the other hand, the combination of other factors/ effects is more important than factor X and one will have to test a different factor, such as X'.

In some textbooks of philosophy of science (e.g., Godfrey-Smith, 2003; Kasser, 2006), the authors indicate that there is some arbitrariness with deriving an underlying overall trend. This was a common misunderstanding by non-scientists. In fact, in science, there are standard procedures and methods that students have to learn during their training. Here are some basic ideas. The most commonly used correlation analysis is the linear fit method which results in two fitting parameters, slope a and intercept b. Its application can be extended to various fitting functions by using different scales. If the linear fit is performed on a linear–linear scale chart, the relationship between X and Y can be written as $Y = aX + b$, and the two parameters a and b are provided by the fitting program. If X and Y are on a logarithmic scale, the relationship is $Y = bX^a$. This gives the power–law relationship between the two quantities with the slope being the power-index a. If X is on a linear scale and Y is on a logarithmic scale, a linear fit gives $Y = b \exp(aX)$, i.e., Y exponentially increases (decreases when a is negative) with X at a rate of a.

278 Philosophy of Science

The underlying trend can also be described mathematically by polynomial functions or sine and cosine functions. If the result is uncertain, one needs to reconsider the variable or to subgroup the dataset. A result with a large number of data points and a large correlation coefficient is considered *robust*; a knowledge-that can be derived based on inductive reasoning from such an analysis. Here, I should mention that a more complicated fitting function introduces more to-be-determined (fitting) parameters in the result. This is equivalent to a proportional dilution of the number of data points among the parameters and hence reducing the robustness or trustworthiness of the result. Therefore, it is better to use the simplest possible fitting function, i.e., a linear fit with two parameters, unless the underlying trend is visually obvious indicating a specific function. How can we determine which fit is better when we have multiple possible fittings? The correlation coefficient is the judge, i.e., the fit with the greatest correlation coefficient is the best one to use.

An appropriate result within the range of data coverage is interpolation in nature, and using the result outside of the data range is extrapolation. There is a potential risk for any extrapolative inference when the prediction is beyond the range of observation. Predicting the future is, in principle, an extrapolation (in time). Prediction, in this case, is not valid from inductive reasoning although it may be made based on knowledge-understanding as discussed in Chapter 9.

Correlation analysis can also be conducted with multiple variables/ factors and is called multiple variance correlation analysis. The requirement is that variables have to be *independent*. If the variables are not independent, a scientist must find the right set. With the availability of "big data", "data mining" software, and "artificial intelligence", multi-variance analyses have become more popular to make predictions. However, I am not yet convinced about their usefulness in science because, if done correctly, these methods may derive "knowledge-that" but not knowledge-understanding. Science is not only about predictions but also understanding. In the topics I have worked on, there are several empirical models derived from data mining, neural network, and artificial intelligence methods. They are not impressive at all although they were developed by groups of renowned mathematicians from famous universities. For example, when asked about specific features in the results, the answers were along the lines of "believe me, this is what the model shows, and the model was developed by famous so and so". The problem, again, is that in science one has to show understanding and not just say "believe me". Reputation and prestige in areas other than the specific science topic do not lend any support to the credibility of a prediction.

Correlation analysis of multiple variables can be performed in terms of orders. One first derives the most pronounced trend or first-order (dominant) effect. Then, one can subtract the first-order trend from the data to obtain the residue from the first-order analysis, as described by John Stuart Mill, and perform the correlation analysis again with the residue to derive the second-order factors that affect the process.

The technicalities of the data collected and used in scientific research are essential to data analysis. If the data used in a data mining or artificial intelligence study are from multiple sources with various qualities, the results can be very doubtful. The knowledge about the data itself is the foundation of scientific inductive reasoning; there is no shortcut here. I hope that the discussion above helps explain the flaws in some theories of philosophy of science concerning Bayesianism.

Polling: Polling is widely used in many disciplines, such as politics, sociology, and psychology. Sometimes, it is presented incorrectly as based on Bayesian theory. Noting what is involved here is often only a probability of occurrence and not conditional probability; it may be better categorized as quantitative stereotyping because usually it is not done by deep-thinking conceptualization, i.e., to derive a law and does not provide knowledge-understanding. The major concerns in these studies have been the sample size and distribution of a poll to avoid sampling bias.

However, to be a valid study, the fundamental assumption in a polling study is that the samples are truthful in telling their opinions/feelings. If there are untruthful samples, probability theory requires that the number of untruthful samples on one side of the question equals the untruthful ones on the other side, so the untruthful samples on the two sides cancel out statistically. This may not be a bad assumption if the issue is not important to the personal interest of the samples. Nevertheless, the researchers have to remember the potential flaws in the method because there is in general no independent way to verify if a sample is truthful or not. Sometimes, if the poll is politically charged with ethical implications, more samples could be untruthful and not in a random way. For example, in one survey, I was asked whether I trusted more a taxi driver or a police officer. Most people have both good and bad experiences with either taxi drivers or police officers. Since this question implies one's political inclination toward laws and order, I am not sure whether the people who do not tell their true opinion are equal on the two choices. I have also observed a few pollings where voters were asked about how they planned to vote on a proposition. An overwhelming number of people said they would vote for the side that was framed as being more politically correct, but this side might be against their personal interests/benefits. However, the voting results showed overwhelming support for the other side, i.e., most people vote with their personal interest and did not tell their true plan of the vote to the pollsters. This told me that simply framing an issue as being on a moral high ground may not necessarily change significantly how people vote if it is against the voters' personal interests. If political science wants to be a science, it has to fundamentally reconsider its theories and methods. A label of Bayesian theory would not make it science.

Probability theory, in general, has to be applied with caution. The current polling method is deeply flawed. On the other hand, Bayesianism, the third application of probability theory explained at the beginning of the section, is mostly irrelevant to science.

10.5 Note on Double-Blind Method versus Traditional Chinese Medicine* (Elective)

Causality always has a special place in philosophy of science. In the theory of science discussed in Chapter 9, causality has to be provided by knowledge-understanding. In principle, the correlation of two observational phenomena does not indicate causality from inductive reasoning. However, very often people demand science to provide causal relationships. In Section 7.5, we discussed causality and found that there are some major ambiguities between potential cause and result, such as the asymmetry (the case of the flagpole), irrelevance (the case of man's inability to become pregnant by taking birth control pills), and a shared cause (thunder and lightning). However, in some cases, there are clear causal relations. For example, a certain medicine has obvious effects to treat a targeted disease or symptom even if we may not have a clear knowledge-understanding of how the medicine cures the disease. For example, in medical science and pharmaceutical science, scientists often do not know, e.g., exactly what chemical compounds in a medicine cure a disease.

The idea of a randomized experiment was first introduced in clinical trials and then later applied to psychology and then education. It is based on the idea of cancelation of various factors if the samples size is large and samples are totally random, as we discussed in the theory of probability in Section 10.4. Although this idea sometimes has been abused, producing some misleading results that confuse society, randomized controlled trials (RCTs) have been applied successfully in medical science and pharmaceutical science. The method is also referred to as double-blind testing and is considered a gold standard in Western medicine and one of the best examples of a randomized experiment based on the probability theory.

The key for double-blind trials to success is "control". After all, any error in medical science and pharmaceutical science can cause death. In a double-blind trial, participants are randomly divided into two groups. One group is given the trial drug and the other a placebo, i.e., a sugar pill, but the participants are not informed about what they are given. These trials have three phases, where each subsequent phase is more serious, under tighter control, and documented more specifically. Using this gold standard, many new drugs have been invented, tested, and successfully applied. The development of the COVID-19 vaccines may be one of the best examples.

I should add a note here: The double-blind method is a scientific procedure and not a scientific method as defined by philosophy of science. Rather, I discuss a case below to demonstrate the limitations of the probability theory. For example, although RCT may be able to successfully confirm an idea, can it disprove one if the result is negative?

In this specific case, the double-blind method is used to test the traditional Chinese medicine (TCM) discussed in Section 9.3. Let me first give a brief introduction to TCM. TCM may be traced back to the 14th–11th century BCE. As I stated in Section 9.3, from the view of philosophy of science, it has two fundamental components: a large-scale observational correlation and a theoretical explanation. The theory is based on energy flow in energy channels inside the human body that Western medicine has not been able to find physiological evidence for. The theory introduces a few story lines to describe the body by the five-element theory (wood, fire, earth, metal, and water) and to explain the connections between the internal organs, people's health, and symptoms. According to the theory, people get sick when they lose internal balance among the different elements. For example, the liver is considered to belong to wood and Spring. It is easy for the liver to lose balance in Spring when plants grow. Water helps wood grow and fire can burn wood. In analogy, the liver could be damaged by furious anger — the fire. The liver has an opening in the eyes and its appearance is in the fingernails. In other words, to learn about the health of one's liver, the doctor would look at the eyes and fingernails of a patient. A damaged liver can be correlated with a yellowing of the eyes and thinning of fingernails. If your nails are suddenly becoming soft and thin or curled, you might want to see your doctor for a possible liver problem, according to TCM. Note that a few thousand years ago there were no blood tests and, within Chinese tradition, performing autopsies was a sin. All information could be obtained by only visual, smelling, verbal, and touching the pulse of a patient. As I explained in Section 9.3, one should not take the "theory" of TCM too seriously or literally because of the potential errors after a few thousand years as new information is continuously discovered. Also, understandably, I think the story line made the profession appear more sophisticated for job security concerns.

The real treasure in TCM is the correlations among food, personal constitution, and the medicinal functions of various herbs as discussed in Section 9.3. It is worth mentioning that traditional Chinese medicine is based on the idea of personalized medicine. According to observations and experiments, humans may be categorized into, say, five different constitutions. Let's call these constitutions hot, warm, neutral, cool, and cold where hot may mean the most energetic and cold the least energetic. For example, the hot constitution is more likely to develop constipation and the cold one diarrhea. Various meats, seafood, vegetables, and fruits were also tested and documented for their positive and negative effects on the health of each constitution. Even cooking methods, such as deep frying, barbequing, baking, steaming, and stewing, could also be an important factor in this mix. For example, people of hot constitution better avoid French fries which may worsen constipation. If this is not complicated enough, gender, age, and season all contribute to this network of knowledge about the health of a person in this empirical database. If one has a chronic health problem, one needs to watch carefully what not to eat and how the food should be prepared according to one's constitution, age, gender, and season.

Now, we go back to the problem of using the double-blind method to test a personalized TCM. Almost all test results conducted according to the double-blind method have been negative so far. How come? Many Western doctors conclude that all Chinese people have been fooled by their doctors! But this is very unlikely because this work was developed over a few thousand years in one of the most populous countries. Some patients were emperors and empresses! If a doctor tried to fool emperors, their life would have been in danger.

In fact, these negative test results may be easy to understand and explain. In TCM theory, personalized medicine is most effective in one of the five constitutions. I recall that a double-blind test is performed disregarding the differences among the personal constitution categories. If personal constitutions are distributed evenly, personalized medicine would work for about 20% of people. Therefore, it is not surprising that, on average, most Chinese medicines cannot pass a double-blind test because they would either be not effective or negatively effective when the test is performed over the whole population. Let us assume five simple idealized medicines, A to E, each of which would work perfectly for one constitute of people, *a* to *e*, respectively, with 1.0 point signifying perfect effectiveness. Let us further assume its effectiveness is reduced by 0.5 points sequentially to the next neighboring constitution, as shown in Table 10.1.

Amongst the five personalized medicines, medicine C is most effective over the entire population with overall effectiveness of 0.4. Depending on how a passing grade is assigned, it is possible that all personalized medicines fail the double-blind tests although each of them, ideally, could work perfectly for each corresponding subgroup. This phenomenon is called the Simpson paradox (1951); the body constitution in this case is called a confounding variable in statistics. The question is how one draws a conclusion from the result.

Table 10.1.   Double-blind tests of personalized medicines.

Medicine	A	B	C	D	E
Constitution *a*	1.0	0.5	0	−0.5	−1.0
Constitution *b*	0.5	1.0	0.5	0	−0.5
Constitution *c*	0	0.5	1.0	0.5	0
Constitution *d*	−0.5	0	0.5	1.0	0.5
Constitution *e*	−1.0	−0.5	0	0.5	1.0
Net test score of medicine	0	1.5	2.0	1.5	0
Normalized effectiveness over whole population	0	0.3	0.4	0.3	0

As I explained, the Chinese medical theory is only *one* of the possible interpretations of the large-scale observations conducted over thousands of years in one of the world's most populous ethnic groups. If it has not fundamentally changed from the theory developed three-thousand years ago, the chance for it to be problematic is large. Just consider Aristotle's theory that was developed a thousand years after TCM. We now know that his theory is mostly wrong. Questioning and challenging TCM *theory* is reasonable and should be encouraged. However, one should not immediately reject the *empirical results* even if the Chinese medical theory is not scientific enough and lacks anatomy evidence. For sure, some of the predictions developed from the theory are not perfect and sometimes the treatments do not work. The energy channels in acupuncture that cannot be found physiologically could be simply considered a placeholder in development according to instrumentalism. I found, from the viewpoint of philosophy of science, that the complete rejection of large-scale and long-term empirical observations, when rejecting the Chinese medical *theory*, is a fundamental flaw in today's popular reasoning in medical science. One can challenge the TCM theory, but should draw a careful distinction between the observational correlations and the theory proposed to explain the observation.

There are two possible conclusions when a double-blind test cannot positively prove some traditional Chinese medical treatments or medicines: Either the Chinese medicine theory is flawed or the double-blind method is not applicable to this situation. In either possibility, one has not falsified the results of the long-term large-database observations. This has a great similarity to the situation with the Michelson–Morley experiment falsifying Galilean relativity, as discussed in Section 6.4. (Today, Galilean relativity remains valid in non-relativistic situations.) As I showed in Table 10.1, the double-blind testing method may have limitations and the experiments conducted to test the TMC were flawed because for personalized medicines one should not expect each of them to be effective for the entire population. For example, with personalized medicine, the participants in a double-blind test have to belong to the targeted subgroup.

In a *Nature* editorial article (2007), it says, "So if traditional Chinese medicine is so great, why hasn't the qualitative study of its outcomes opened the door to a flood of cures?... it is largely just pseudoscience..." I found this comment not sensible. The Editor clearly has confused two things: the observed correlations and the explanations. They concluded Chinese medicine to be an outright pseudoscience based on some strange reasoning: If some field of research has not yet "opened the door to a flood of" improvement, it is pseudoscience. This is a much stricter requirement for science than even Popper's demarcation. Many new science fields would be pseudoscience according to this reasoning. The reason I cite this article here is that many naïve medical professionals have made arguments and story lines according to this article to dismiss the scientific significance of TCM entirely. In extreme cases, some medical professionals have given lectures and seminars describing TCM as a medical fraud and scam.

An unbiased scientist would question the explanations or theory of traditional Chinese medicine but would have little reason to doubt the overall observed correlations. The application of a double-blind methodology to testing personalized medicine will have to go to the subgroup level for further testing. The categorical dismissal of traditional Chinese medicine techniques makes people wonder whether the patients and doctors in China, over the past few thousands of years, did not know anything about their health. If the treatments and medicines that people received had not worked over a few thousand years, people would not have seen their doctors. Medical doctors would not have survived as a most respectable profession in Chinese society for thousands of years. Plagues did not develop into rampaging pandemics in China, although there have been a few localized epidemics in a thousand years. These localized epidemics killed a much smaller fraction of the population than the pandemics in

Europe, though China might be closer to the sources of the European pandemics. Traditional Chinese medical theory, although with flaws, must have some validity.

The discussion of traditional Chinese medicine reveals that the gold standard of the double-blind tests has its limitations when the subject depends on multiple factors, especially when the results are negative. When testing it for a targeted subgroup, the participants should be limited to the targeted subgroup.

10.6 New Riddle of Induction* (Elective)

The objective of this riddle is to show that it is impossible to develop a deductive theory for the confirmation of inductive observation discussed in Section 4.4. To start this riddle, we recall the deductive logic "all B is A and all C is B, therefore, all C is A". This conclusion depends only on the "form" of the logic and not on the specific definitions or contents of A, B, and C in the case shown in the left panel of Figure 4.1. Logical empiricists wanted to show that inductive logic can be presented in the same way, so that conclusion depends only on the form and not the content of the statements. The riddle disproves the possibility.

The riddle goes like this:

One observer has observed many emeralds which are all green. The observer makes the following argument:

Argument 1: All the many emeralds that have been observed, in many diverse circumstances prior to 2050 AD, have been green; therefore, all emeralds are green.

Another observer also made the same observation but has defined the color of emeralds to be "grue" which stands for a color between the colors green and blue. They further define "grue" as green if it was first observed before 2050 AD and as blue if not first observed before 2050 AD. According to this, the second observer makes the following argument:

Argument 2: All the many emeralds that have been observed, in many diverse circumstances prior to 2050 AD, have been grue; therefore, all emeralds are grue.

Because arguments 1 and 2 both have the same *deductive form* as a syllogism, we have the following:

> Argument 3: All the many E's (for emeralds) observed, in many diverse circumstances prior to 2050 AD, have been G (for green/grue); therefore, all E's are G, where G is defined as green if it was first observed before 2050 AD, and as blue if not first observed before 2050 AD.

Both arguments 1 and 2 are drawn based on observations and hence are inductive logic although they define the color differently and both generalizations are to the future which has problem with induction. The predictions by both arguments are the same before 2050. However, after that, according to argument 1, all emeralds are green, but to argument 2, all are blue. Argument 3 is drawn from deductive logic (note that this is using the general form of logic) to make a prediction based on the fact that both arguments 1 and 2 have the exact same "form" in terms of deductive logic. But the conclusion is that after 2050, the emeralds have color G, which is green or blue. The logical empiricist effort to make inductive logic deductive logic failed. The "new riddle" concludes that it is impossible to make inductive inferences deductive to produce a justifiable conclusion.

This new riddle has generated many debates in philosophy of science and is considered to be the final nail in the coffin for inductive inferences. It also announced the failure of logical empiricism's rational reconstruction of science according to the theory of language and logic. The consequential conclusion is that inductive inference is unscientific and cannot be used in science because its predictions could be wrong. This conclusion about inductive inference is obviously problematic because science critically depends on observations and experiments as well as their generalization. However, there is another possibility — the logicians, i.e., the constructor and examiners of the riddle, could be wrong.[3] In this case, inductive reasoning may still be fine in science. Therefore, the key question is whether there is a logical flaw in the new riddle or not. If yes, where does the error occur? However, if yes, the conclusion drawn from the riddle is not valid.

---

[3] About this possibility, recall that a fundamental question for a scientist to ask is, "how do I know I am not wrong?" They should have asked this question to themselves.

Let me first comment on the pick of the colors used by the author of the new riddle, Nelson Goodman (1983). People see color by the light reflected in their eyes from the subject. In science, we have learned that the strongest solar radiation is in wavelengths of the visible band that peaks at the color green. Human eyes, through their development, have adapted to this color, so that it is the wavelength we are most sensitive to. Other animals may have adapted to other wavebands. If you pay attention to the color of traffic lights, you will find that the green light appears to be colored more diversely than other colors, from a yellowish-green to a blueish-green, and to being outright blue. Therefore, Goodman likely picked his color carefully to make the riddle more confusing. For example, if a new technology produces the most reliable and brightest blueish-green bulbs that use the least amount of energy at the lowest prices, most traffic lights would use this technology. However, if this new green light is more blueish than greenish, potentially someday in the future, say 2050, we may make a change in our language to refer to this "green light" as "blue light". Therefore, some philosophers think green, blue, or grue is all *semantics* and there is no logical problem in the new riddle of induction.

Now, let us ignore the observation/induction part of the argument and focus on the conclusion of Argument 2. We change the example of color to a simple arithmetic example, in order to avoid Goodman's trick or the potential semantic issue; the issue may be understood more clearly. Let us define a number "fove" which represents a number *between* numbers four *and* five; this is equivalence to color grue being between colors green and blue. Think of argument 1 as "$2 + 2 = 4$ (four)" before and after 2050 AD. According to Argument 2, i.e., "$2 + 2 =$ fove", this number fove is 4 before 2050 AD and is 5 after 2050 AD. Argument 2 would be correct prior to 2050 AD because "fove = 4". However, after 2050 AD, it is "5" so that $2 + 2 =$ fove $= 5$ after 2050 AD. Arguments 1 and 2 indeed have the exact same form, but after 2050 their results are different because of a bug Goodman placed in argument 2 by having a definition change with time. When introducing it, the color "grue" is a color *between* green *and* blue, but when using it, the grue is *either* green *or* blue depending on the time of observation. In our arithmetic example, fove equals *either* 4 *or* 5. Argument 2 changes the definition of the color without changing the naming of the whole color system. This is a logical flaw! In our example of number fove, while one has the right to redefine the symbol and pronunciation of a number, the whole arithmetic symbol system has to be changed accordingly at the same time. This means that the swap between 4 and 5 has to be made in 2050 AD in the whole society on everything using 4 and 5 including all the math books, accounting, finance, money system, address numbering, and all past documents. This cannot be simply referred to as a "semantic" or linguistic issue as some conventional theories of philosophy of science have argued, although it is true that to refer to the color of emeralds as green or blue may be more a semantic or linguistic issue. The flaw in the new riddle is now obvious in the arithmetic example. It is to change from "between-and" to "either-or" as I highlighted above. The year 2050 was used to distract attention. And, he succeeded. I wish the logicians and linguists were not fooled by the trick.

Therefore, the new riddle contains a flaw in its logic. It cannot be used to question the validity of the inductive inference. In fact, it cannot prove or disprove anything. Although the new riddle helped identify the flaws in logical positivism, it has been focused on the "form" of logic. In science, we focus on "reasoning" and not merely its form. The conclusion of the discussion about the new riddle should have been that the form of the logic is not foolproof because one can insert logical flaws into the contents of the argument as Goodman did and that using symbolic presentation in philosophy of science is fallible. Unfortunately, people have spent too much time and effort to help solve the new riddle instead of pointing out its logical flaw. Its conclusion that inductive inference is fallible for science is again a fundamental mistake by many conventional theories of philosophy of science as I pointed out earlier.

From the new riddle, it is clear that the conventional theories of philosophy of science have gone on a wrong path when people debate heatedly about "grue" and "bleen" (for blue-green). If the color is defined in terms of wavelengths that cannot change with time, there would be no ambiguity to argue about. This is where the requirement of evaluation in science can help. Focusing on pure logic/linguistics, such as the "form", was misdirected in the early years of the development of philosophy of science even if it provided some interesting perspectives. After all, if a result critically depends on linguistic tricks, it is not a real scientific result. The new riddle may indeed have some educational value of logic in this regard. However, this is not relevant to scientific reasoning and could mislead students.

The problem that *New Riddle of Induction* reveals is the following according to the theory developed in this book. The riddle raised a correct question concerning simply using deductive reasoning to "generalize". As we discussed before, deductive reasoning cannot generalize. Therefore, to generalize, one has to use inductive reasoning. However, it is known that induction is potentially flawed. Therefore, simple induction cannot be used for scientific generalization which is used as the start point for deductive prediction. The issue here is how to generalize without potential fallacy. The existing theories of philosophy of science oversimplified the process of making a prediction based on induction. It actually includes two steps: generalization from limited observations by induction and making a prediction by deduction from the conclusion of generalization. Whether the prediction is not fallible depends on the quality of the conclusion of the generalization. If the generalization results in a law of nature or understanding, the prediction can be robust within the limit of the validity of the law and understanding. Otherwise, the major potential fallacy for (deductive) prediction is the garbage-in-garbage-out. Some theories, such as covering law theory, proposed that prediction needs to be based on a "law-like" inductive result. This, in theory, would strengthen the prediction. However, it is not clear what qualifies as a "law-like" observation. Doesn't the fact that the Sun rises every morning qualify for it? Although there has been no exception found so far, Hume still doubted its prediction power. Does "people prefer immediate and certain consumption to future and uncertain one" qualify for a law? As we discussed in Section 7.6, the argument can be flawed although it is based on observation. The conclusion is then that prediction has to be based on laws of nature and/or knowledge-understanding.

10.7 Roles of Philosophy of Science in Future Science

Although we have discussed the traditional philosophy of science in Section 2.6 and its successes as well as failures in Section 8.7, people have different ideas about it as indicated by various definitions of philosophy of science. I prefer the definition given by *Philosophy of Science: The Central Issues* (1998) which is, "Philosophy of science is the investigation of philosophical questions that arise from reflecting science". These philosophical questions have to be in common among all, or at least many, science disciplines and cannot be answered by the knowledge gained from an individual discipline. For future studies of philosophy of science, it should include a full theoretical investigation of problems, such as the potential causes of flaws in scientific reasoning, the reliability of scientific theories, the relationship between science and truth, the relationship between mathematics and other branches of science, how to make sound scientific judgments with insufficient information, how science reconciles conflicting ideas, and how to expand the concepts of science into new fields of research.

The last subject is the most urgent. Over the last few decades, many fields of research have started adopting concepts of science, such as psychology, economics, medicine, sociology, archeology, and, to some degree, computer science. Each of these fields has accumulated large amounts of information and data and is still accumulating more. As I have previously mentioned, I have not seen an overarching theoretical framework, conceptualization, or a solid paradigm forming in each of these fields. I would call for more participation from scientists in the construction of a new theory of philosophy of science that provides a structure and theories that can be applied to these fields plus the whole of science.

The ultimate goal of science is to invent new knowledge in order to deepen our understanding of the natural world. The new knowledge has to lead us closer to the truth with a minimal possibility of making systematic mistakes. Philosophy of science may need to study how science is conducted at the individual, research group, and societal levels. Philosophy of science should try to answer the two central questions concerning scientists: "How to get it right?" "How to know I am not wrong?" I tried to answer these two questions in the earlier chapters and will discuss the second question in more detail in Chapter 11. How does the system of science ensure the reliability of scientific outputs while not destroying the system itself and keeping the independent thinking of each individual scientist? I described in Chapter 8 how science is organized to avoid this destruction from happening. This may not be the most rational system, but this is how science has successfully operated for centuries. Is there a better system for future science?

Three key elements of science (i.e., observation, conceptualization, and evaluation) have been identified and the four forms of scientific reasoning (deductive, inductive, dialectical, and Occam's razor) are also identified. The theory of science described is still very preliminary and needs much further development. Furthermore, the theory is based on the problem of knowledge in philosophy. New science disciplines, such as psychology, sociology, economics, and political science, primarily concern the problem of conduct in philosophy, although there is a fraction concerning the problem of knowledge. I recall that the problem of knowledge is mostly deterministic, or at least probabilistically deterministic, regardless of whether we like its consequence or not. Scientists spend much more time verifying that a conclusion is unique and not wrong. The problem of conduct concerns multiple options that an individual can choose from. In this case, the individual's value system, preference, and psychology can be involved. Since this is a matter influenced by opinion, there is no correct choice. Even the "better choice" may change from time to time depending on the person's psychological state and political environment. Paradigms for these sciences either do not exist or are in development. Whether the theory described in this book can be applied to these sciences, especially for the subdisciplines that study and predict mostly human behaviors, is not clear.

A good, experienced scientist must think from a philosophical level by ignoring technicalities. This is what Einstein referred to as seeing the forest. In this process, one would be able to better identify gaps or flaws within scientific reasoning more easily. Philosophy of science should strengthen the general ideas in this process. Without philosophical thinking, when encountering challenges, a scientist may tend to explain the problem by showing mathematics, computer algorithms, and data details, which may overwhelm a challenger. This behavior in science is often referred to as hiding behind the technicalities. Instead, a good scientist can explain the problem and answer the challenging questions more on a philosophical level, such as by using analogy to explain (noting that analogy is not used to prove) the concept and understanding. The ability to explain complicated ideas in a less complicated manner is usually characteristic of a good scientist who has spent more time thinking through these problems more deeply. Therefore, learning to think on a philosophical level is an important part of education and, especially, in Ph.D. training. In this process, one must constantly question one's own ideas. One's scientific adviser and fellow scientists are the best sources for criticism and improvement. Eventually, students should be able to teach their professors with their new findings in the program. Then, they are ready to graduate! Philosophy of science should theorize this process.

Another immediate urgent task for philosophy of science is to clean up the false concepts and ideas developed in the conventional theories of philosophy of science, especially the concepts of scientific methods and logical reconstruction of the science. These flawed ideas have now penetrated every fiber of society. I have heard from many scientists to explain about scientific management at the national level that has required scientists to follow the failed H-D method offered by the conventional theories of philosophy of science.

In philosophy of science, there is a theory called reductionism which thinks that all social science problems can be reduced to psychology which can be reduced to biology which in turn can be reduced to chemistry and then to physics. As a physicist, I think this a flawed logic. It is true that everything, including living things, comprises elementary particles. In a living thing, every particle is undergoing constant changes. There are practically infinite possible changes. In physics, the simplest situation of a large number of particles in motion is an ideal gas. To describe an ideal gas, one may employ statistical mechanics, an approach that involves complicated mathematics. When a collection of particles is not uniform and not gaseous or solid but gluey and alive, there is no existing general physical theory that can fully describe the process, not mentioning that the volume of the collection is infinitely large, and the particles can have electric charges. I do not think it productive to discuss the question now and maybe in the next one thousand years. Any serious theory of philosophy of science should not be based on reductionism.

Here, I should add a side note that the new knowledge does not need to be "significant" (revolutionary) knowledge that conventional theories of philosophy of science like to discuss. Whether a new piece of knowledge is significant or not should not be measured by philosophy of science. The efficiency of the return from the investment of their time is a question for individual scientists or science programs to consider. Therefore, to philosophy of science, the issue is only to get it right, per se.

Questions for Thinking

1. One theory of philosophy of science presents Darwin's principle of natural selection (PNS) as "given two competing populations, X and Y, if X better fits the environment than Y, then, in the long run, X will leave more offspring than Y". According to the author, PNS assumes its result is true and is, hence, somewhat circular. Darwin's theory is therefore problematic. But, is this a problem in Darwin's theory or in the author's argument? Hint: Darwin's theory should have been described as follows. There are two possibilities X and Y. (Since you have observed only X and not Y, X fits better. The author misrepresented PNS as a small step, which is deductive, in PNS.)

2. One theory of sociology of science refers to "science as a process". According to the theory, science evolves in a similar manner to an evolutionary biological process. Then, there is the question of how to define what is "fittest". If the fittest is defined as the one that produces the most offspring, an argument can be made that humans are the least fit because a couple produces only a few offspring in their whole life. In contrast, mosquitos fit better than humans. They not only produce a lot more offspring in a very short life, but also deploy a better strategy by outsourcing the food growing, preparation, and digestion to their prey — humans. They need only to suck out the most delicious nutritious food — blood — from humans. Do you think that science is a Darwinian process?

Chapter 11

Challenging the Existing Knowledge

Scientists have to either discover or invent new knowledge that does not exist or show that existing knowledge is false. The knowledge discovered through new observations may be the least controversial. For example, Leeuwenhoek's invention of the microscope brought the new ability to observe small creatures, although human beings have suspected for some time that bugs in food could be the cause of some illnesses. The Hubble Space Telescope shows new knowledge in distant space. Although there are fewer disputes about these new observational discoveries, at this level, they are all knowledge-that! Science is more about deriving new knowledge-understanding, which requires explanations and interpretations, especially quantitative ones, of the new observations. New knowledge-understanding would have to challenge the existing one.

To challenge the existing knowledge, a scientist must first have the courage to challenge the existing ideas, which is part of the profession for every experienced good scientist. My observation indicates that this courage is nearly independent of personality or gender. This is understandable because one may have spent ten, twenty, thirty, or forty years to reach that moment. During this long period of time, even if one was originally a shy and quiet person, one could have cultivated this courage through many smaller-scale debates for this moment. Based on my observations, some scientists who are quiet and less argumentative are actually the most courageous. With courage, any scientist, whether they are an experienced debater or not, does not want to make a false claim, which can damage one's reputation. How can one prevent making false claims?

11.1 Sanity Check

When you have derived something new or you think you have proven existing knowledge to be wrong, you need first to have a "sanity check". This means that you should pause and jump out from the details in order to examine your work on a philosophical level since you should now know how useful and important philosophy and philosophy of science are. You have learned about seeing the "forest". You should ask yourself the following: Why haven't other people invented it? Why *not*?! Can the new knowledge be used *elsewhere*?

You may have a few reasons for your answers to the questions, but "because I am smarter and more creative than them" cannot be among them. You may find my statement very strange because you may have never thought of yourself as being smarter and/or more creative than others. If you do not think of yourself as being smarter than others, you are unlikely ready to challenge an existing theory. Challenging existing knowledge requires tremendous self-confidence, where you do need to "feel" as if you were smarter than anyone else. Otherwise, you would unlikely be able to follow through the falsification process when everyone is criticizing and attacking your new invention from every possible direction. However, I said that being smarter and more creative should not be a reason for *you* to invent new knowledge. Why is that? In reality, knowledge is accumulated by human beings all over the world over a few thousand years of civilization. Now, there are a few Nobel Prize awards every year, where each of the awardees is definitely smarter and more creative than you. If you disagree, think of scientific giants like Newton, Einstein, and Darwin. Therefore, one needs to be self-confident while not being overconfident. Without self-confidence, one would not be able to defend the idea during the falsification process of a major invention which falsifies existing knowledge. Overconfidence can result in major mistakes or flaws.

Throughout the course of the investigation that leads to the invention of new knowledge, you likely had many nights when you could not fall asleep either because of frustration or excitement. During these nights, you must have had many (not just one!) great ideas. Some of these ideas did not work out, but others were modified and did work out. You must have thought each of these great ideas essential to the new invention. However, being intelligent or creative alone is not sufficient for someone to invent new knowledge. At the philosophical level, I am asking you to use a single sentence to explain why other people did not think the way you did. The answer should be among what you have learned about philosophy of science in this book! There are only three reasons that are clearly acceptable: new observations and/or new conceptualization and/or a new evaluation method or theory. You need to be able to trace the origin of them since the new idea could have been invented by others.

The invention of an idea may have diverse possible sources. However, in science, new observations are more often the trigger of new ideas. A "new observation" is associated with identifiable new capabilities, new facilities, and/or new methodologies. You should be able to name it and explain what the newness is. If you find that you have a hard time naming it, you may want to lower the level of your claim. This is because the complexity of our present knowledge makes it difficult to make a top-level invention based on homemade devices or small laboratories, except in yet to be developed fields of investigation. The good news is that many national and international projects have developed open-source data policies so that everyone can have access to new observational or experimental results with only minimal time delays.

The invention can also be based on existing observations of different phenomena. In this case, the invention makes a connection among these phenomena. In science, we often use the analogy of a blind person feeling an elephant for this type of invention. Piecing together many observations is conceptualization. For example, Newton's theory was not caused by a falling apple hitting his head. Rather, he took the gravity experiment and concept of inertia from Galileo plus observations and mathematical descriptions of the heliocentric motion of the planets by Kepler to conceptualize Newton's laws of motion and the universal law of gravitation.

If the reason identified is something other than the three previously mentioned reasons, i.c., new conceptualization and/or new observation and/or a new evaluation method or theory, one has to be extremely careful! For example, if the reason is that you have identified an error in an existing theory, you may need to think harder. This is because if an important theory has an error, it would be inconsistent with some observations.

If no contradiction has been reported and hence studied, there is a chance that the theory is not important. In retrospect, the theoretical model I questioned during my Ph.D., as described in Section 5.3, concerns the physics of a subdiscipline level. The new knowledge derived from that effort was based on new observations. In this case, the new capability of the new observation was a new satellite equipped with better instrumentation and launched into a new region in space. Eventually, Southwood and Kivelson cut the connection between the model and the upper-level paradigm. As a result, the challenge to the model is no longer a challenge to the paradigm.

11.2 What Is New and How Did You Do It?

You also need to ask yourself the following questions: What is new? So what? Who cares? The answers to the first question should be in less than three sentences without using any equations or acronyms. If you are unable to answer it in less than three sentences, you need to think more carefully about what exactly is new. In doing so, you will find that most of the technical details that you have been excited about are NOT important. The actual number of sentences is an indicator of the importance of your new idea. For example, "I have invented a new drug to cure AIDS." With this single sentence, everyone understands the importance of your work. The problem is then whether you really achieved what you claim. If you were the first to observe gravitational waves, you would find that many people do not know what a gravitational wave is. You would need to add a sentence such as "according to Einstein's general relativity..." before making your claim on the first observation of gravitational waves. If you have only contributed to the grand effort of the observation, you may need to add a third sentence that specifies your contribution. If you only contributed to a peripheral component in this effort, you need to moderate your claim accordingly.

The "so what" question concerns the significance and consequence(s) of your invention. This also concerns your direct contribution — not the overall grand idea: whether your contribution is an essential part of the project or whether it is peripheral. This thinking exercise is to help one consciously recognize their role. Do not claim something that does not belong to you. Here, I should mention that in one's science career it is impossible for one to always play the key role for every research project. With teamwork, one should work as hard as possible even when playing a supporting role.

The "who cares" question concerns the potential applications of your invention. Often, this is about future financial prospects. If someone cares, they will financially support the continuation of the project. If the new idea is a fundamental science and has no specific interest group, then you need to plan for a long period of time without a significant increase in support even if you think the invention is earthshaking.

These three questions would help a scientist draw a clear picture of the idea's position in society. One may argue that in this case, good salesmanship may help the process. But I have to warn you that the answers to these questions are reviewed within the science community. Scientists, in general, do not trust the salesmanship of businesses and commercials. Any exaggeration would be received negatively; a label of a salesperson may follow a person in their whole career. Therefore, one has to think carefully to accurately phrase their answers in a defendable manner. These three questions — "What is new?"; "So what?"; and "Who cares" — are questions that most experienced scientists ask when they are reviewing a science manuscript or proposal.

Next, while we are still at the philosophical level, we provide some technical specifics. First, we try to substantiate the "newness" aspect of the sanity checks. As we have identified three key elements in science, one has to be able to pinpoint which of these three key elements is new.

Observation: You have been able to identify a new observation, facility, and/or methodology that led to the invention in the sanity check. We now ask you to identify exactly what new information was not available or unknown before. If no new information is introduced, no new unknown can be derived purely from observation. If a new known quantity that was previously unavailable is now derived, either an additional measurement or an assumption must have been made. Is it due to new parameters, better spatial and/or temporal resolution, higher accuracy/smaller uncertainty, a new processing algorithm, or a new environment/parameter range?

Very often, because of the complicated technical issues involved, it would not be straightforward to pinpoint whether a new piece of information is from direct measurements or assumptions. In this case, it would be a useful practice to count the total number of pieces of information and the total number of answers. Each measured parameter or each mathematical expression invoked is counted as a piece of information. Here is an example. I remember a seminar I attended a few years ago on a new method of observing a black hole. The speaker claimed to be able to enhance the spatial resolution of the image by a factor of 100 of the black hole. This was very remarkable and impressive; this means that a single point, i.e., a pixel on an image, became an image of 10 × 10 pixels. The speaker showed an image of black hole with many details. I then asked him where the information for this increased resolution comes from. Is it by interpolation? "No", he replied. It was not from an interpolation but from his complicated new algorithm. His new algorithm was mathematically advanced, and with this, he reported that he was able to resolve the surface features of a black hole. I asked whether the data are still from the same instrument. The answer was yes. It is common knowledge that a black hole on a telescope image is a single pixel because it is very far away. Now, he could show it as 10 × 10 pixels while not using interpolation. Where the additional 99 pieces of information come from needs an explanation; the speaker could not answer this obvious question. He struggled and eventually, it became clear that he used 100 images, which were taken over a long period of time of the same black hole. He took a single point from each image. The light intensity of the 100 measurements becomes time-series data. He assumed a model of a black hole with some surface features to construct a 10 × 10 image. He then converted the total light intensity of the black hole model from this 10 × 10 image into a single-point light source. He then allowed the black hole rotating to produce 100 images of this single light source to produce time-series data. He adjusted the surface feature distribution and rotation in the model to compare model output with the 100 observed light intensities of the black hole so that the time-series data appear consistent, to some degrees.

Now one can see, in science, fact, observation, interpretation, and evidence may not be simply related. He then concluded that (1) his model of a black hole is valid and (2) the black hole has surface features and rotates at the same rate as his model when the model prediction and the observation matched. I recall that I discussed this reasoning, in Section 7.2 on abductive reasoning, as using a single equation to solve two unknowns. As we know, abductive reasoning is fallible. The major potential problem is that multiple models can produce the same net intensity variation over time, of a span of a few years; for example, the spin axis of the black hole can be in many possible directions. A surface distribution can be derived for each tilt angle. It assumes that the black hole is a sphere and all variations are due to the rotation of its surface. Of course, the assumed surface features are only one of the possibilities. However, there may be structures and materials between the Earth and the black hole; one cannot distinguish the features on the surface from those not on the surface during the period when the 100 images were taken. In retrospect, if the speaker had learned philosophy of science, he would have considered more consciously these possibilities and, at least, would have been better prepared to answer my questions more quickly and more concisely.

Conceptualization: Conceptualization is high-level inductive reasoning from specific to general of multiple pieces of information. A new conceptual model is often triggered by new observations which the existing models may have difficulty explaining. When challenging an existing model, the first thing to do is to compare the paradigm at each level to understand at which level the paradigm of the new idea diverges from the existing one. You must be able to pinpoint the exact divergence point and provide convincing arguments for your alternative approach. You will not be able to avoid this question because the proposer of the existing model is there to defend their model! These arguments may include but not be limited by the conditions/assumptions in each paradigm, key features of observation and mathematics of governing equations (with the included dominant effects, initial conditions, and boundary conditions). For each paradigm, there is a set of approximations. The condition and validity of the new assumptions have to be carefully examined. The dominant effect can often be identified in the governing equation set. To simplify the problem, a subset of the equations can be used. However, this may be where some important effects are overlooked. Sometimes, it is difficult to pinpoint an effect. "Instabilities", for example, are widely used as a rescuer.

In some research fields, unspecified "instabilities" are equivalent to saying "something I do not understand". They can become a liability that may later haunt the proposer.

Evaluation: There are two types of evaluation. The first is the processing and analysis of observational data. The common problems in data analyses have been discussed in Sections 9.3 and 10.4. Many studies start with case studies of a few carefully selected examples that include the "wanted" features clearly visible and then are followed by statistical studies. More often, these examples are extraordinary cases based on what the investigation wants to conclude, which may sometimes carry biases. It is important that presented examples show the properties and values around the averaged values of the whole database, i.e., in the middle range of the statistical analysis. If the values of the examples are at the extreme of the statistics, the relationship between the case study and the statistical result is suspicious.

Conceptualization also needs to be evaluated. This is THE most important feature of science, which distinguishes science from other types of investigations. Some of the science disciplines may not have mathematically based theories but have computer models that are widely used, for example, in economics, to make descriptions or predictions. A frequent problem is that the approximations and theoretical foundations of such a numerical/computer model have not been carefully investigated. The reliability of the predictions is basically unknown. The predictions can be reasonably good at some time, but poor at other times, and at best if not totally misleading.

In one of the *science* projects that I participated in, a team took an *engineering* approach. They divided the overall science problem into several components, recalling Descartes' divide and conquer method, using a numerical simulation code for each. They assumed that the science problem can be simulated with the serially coupled codes. Therefore, the project in this approach became to write a software interface between each pair of connecting codes. So, they found the most talented programmer for each interface. However, at the end of their simulation, the results were irrelevant to the observation. The difference between their prediction and observation is as large as a few orders of magnitude with an opposite tendency. There were a few fatal mistakes with this approach. The first and most important flaw is that each of the codes was developed based on a specific theoretical model, where each model is based on a set of assumptions. However, there may be intrinsic conflicts in the assumptions among different models involved. Even though the interfaces can smooth and match the solutions from two neighboring codes numerically and pass the information from one code to the other, scientifically, the final solution may not be related to reality. This problem occurs widely in many scientific fields. In some cases, when the output is inconsistent with observation, artificial procedures are introduced to make the results have some resemblance to observation. Then, the code would be considered "working". This situation is reminiscent of the monster Copernicus referred to when discussing the geocentric model in Section 1.4. I think using serially coupled models with inconsistent assumptions is potentially cancer in some disciplines of science.

In many science problems, "critical points" exist. At these critical points, the controlling processes do not change in a continuous manner. A famous example is Parker's solar wind solution as discussed in Section 6.4. When the boundary condition changes slightly, the solution becomes completely different. Although mathematically pretty and sound, the solar breeze model could not predict the existence of the solar wind. A critical point may mathematically correspond to a singularity in a solution, at which the solution jumps discontinuously such as when crossing a shock or initiation of an avalanche. Some numerical models that describe the continuous processes may not be able to correctly predict discontinuous processes. A stock market crash, for example, is a critical point that is widely and easily recognized, only afterward. Until people understand the causes of discontinuous processes, such as with the stock market, numerical models will continuously make false predictions near the critical point, which can lead to false alarms or a missing forecast of a crash.

11.3 What are Acceptable Proofs?

What is considered an acceptable proof depends on the convention of a science discipline. Different fields accept different degrees of concreteness of a proof.

Commonly acceptable proof: There are three basic types of acceptable proofs. The first type is a mathematical proof if there is one. This mostly concerns theoretical models. Because all mathematically based theoretical models hold only under the conditions in which involved laws are valid, the applicability of the theory has to be carefully examined. For example, Newtonian mechanics is only valid when the speed of an object is much smaller than the speed of light. However, this also depends on the requirements of the application. In some cases, a relatively small speed can produce an error that is large enough to affect the application of the results. For example, the speed of a satellite orbiting Earth is a few kilometers per second, which is much smaller than the speed of light at three hundred thousand kilometers per second. However, when using a global positioning system (GPS) to accurately determine a location on the surface of the Earth, the effect of relativity must be accounted for. Therefore, before a mathematical proof (observational), validity for all assumptions/approximations has to be verified and tracked throughout the proof.

The second type of proof comes from observations. To untrained eyes, the data may often appear as random noise compared to theoretical calculations. When one stares at data for a long time, it is common that more features can be recognized. The familiarity with data could be a good thing that requires a set of mentality and skills, but it can also be a bad thing because one can become biased. Information carried by observation is often condensed into figures. For a result with high impact, the main features have to be obvious to *untrained* eyes. If they are not, one may have to find a different way to present. If there is no way to present the features convincingly, the features may be imagined.

The selection criteria for a specific phenomenon must be written down unambiguously because one could find that the criteria drift when more cases are examined. The list of criteria needs to be examined without logical flaws and the data reduction procedure should also be written down unambiguously with documented justification for each step. Now, more studies have used computer programs to select cases. The good news is that computerized criteria are less ambiguous and do not change from case to case. However, the algorithm has to be carefully tested with the database so that the criteria do not falsely select or reject some events because of special situations of these events. Researchers should manually (visually) inspect a significant number of examples. I have seen many problematic event selection algorithms.

The third acceptable proof comes from citations as discussed in Section 7.2. This, for citations to be considered as proof, may be surprising but it may be very important in the referee process. A scientific study comprises many mini-steps, where the reasoning of every step may be under a microscope and subject to falsification. Very often, reasoning is needed to produce a link between observation and conceptualization in these mini-steps. You may need to cite theoretical and/or observational works to make these connections. The connections need to be shown either in the new work or previous works which need to be cited from *peer-reviewed* journals. The cited points have to be clearly shown by either data or mathematics as discussed above or clearly stated in the abstract or conclusions, i.e., not only hidden in passing text. The logic here is that these cited works have been reviewed by experts who are as good as the referees of your new work. Therefore, the referees cannot challenge these published works during the referee process. If the referee wants to challenge a published work, they will have to write a separate paper that has to go through its own referee process. Therefore, citations are an important mechanism in science to fill the gaps in reasoning, in addition to providing inspiration for the new idea. Although a referee may tend to request the author to cite their own works, the self-promotion by the referee is unlikely to make an unimportant work important. With explosions of information, people have been spending less and less time reading original works. This is more of a problem. As a result, in science, often some relevant caveats in the text of cited works are buried and are forgotten by later works.

People tend to use the narratives that circle in a field to defend their works. Sometimes, the narratives can be different from the original idea and do not prove the point being cited. For example, there is a famous experiment called "visual cliff study". In this experiment, 6-month-old to 14-month-old infants were put in an environmental setting that is safe but creates a cliff visually. The infants could crawl into it for some incentives, but most infants refused to crawl into it. This experiment showed that humans have developed the concept of depth or height by that age. The authors of the study specifically stated that "This experiment does not prove that the human infant's perception and avoidance of the cliff are innate". Nevertheless, this experiment has been widely cited as evidence that "humans innately fear falling from a cliff" or "humans have an innate fear of heights", a misquotation of the original work. This is an extremely problematic case for a discipline that wants to be called science.

The problem with citations is that by themselves, i.e., without other acceptable proofs, citations alone are unable to prove a new idea/understanding because if the idea has already been published before, it is no longer a new idea in science. One has a sanity check problem.

Possible proof: Again, this is highly dependent on the level of development of a field.

With the advancement of computer technologies, computer simulations have become a significant tool for scientific research. A first-principle simulation model numerically solves a set of governing equations to provide a solution for each variable in space and change with time. A simulation can produce a nice animated movie to convince people that simulation describes what is happening in reality, at least in idealized situations. However, one has to first distinguish whether the simulation is used to evaluate a problem for engineering purposes or to prove a new idea for scientific purposes. In science, we care only whether simulation provides additional information or proof for new ideas. As we discussed in Section 9.2, numerical simulations are known to introduce leakage and contamination in deductive reasoning. Sometimes, these unwanted effects can dominate the solutions.

Once, a scientist passionately argued with me that the artificially introduced effects in his code are not only the numerical effects by themselves but are needed to stabilize the code. Therefore, the numerical artifacts are absolutely necessary and *factual in nature*, he stated. In order to stabilize the code, he exaggerated the artificial effect by a factor of 1000 times so that this effect is numerically greater than the main physical terms! Essentially, he was simulating something else and not the processes he claimed to simulate.

In the early developmental stages of simulation models, these numerical effects were unknown or no problem to worry about. In the case of space physics, the problem of numerical artifacts in simulations only became widely known in the 1990s. In many other disciplines and subdisciplines, these problems are not fully appreciated yet. I still see presentations of new simulation results that are dominated by artificial effects. Over the last few decades, we have accumulated a great deal of knowledge about artificial effects and the capability to recognize them in simulations.

A computer simulation model is developed according to a certain set of assumptions which are included in the governing equations plus initial and boundary conditions. These assumptions have to be carefully examined in order to be applicable and used to simulate a real problem. For example, I saw a presentation of the evolution of the universe with a large number of randomly moving particles. It shows that these particles collide and coalesce into some structures that are supposed to be galaxies. However, when I asked about the equations or mechanisms governing their motion and collision process (e.g., if two particles collide, are they broken into pieces or attached to each other?), the presenter could not understand the essence of this question. To me, this is a clear case that this simulation was not a first-principle simulation code and not developed under the guidance of a reasonable paradigm. Therefore, the simulation is irrelevant to the process it was supposed to describe. A simulation code has to be tested first with simple conditions under which analytical solutions are known. Then, there is a set of standard tests that needs to be applied, such as convergence tests. A new numerical code or model has to be validated independently with observational verifications under *relevant* conditions before it can be used to simulate a complicated problem of interest.

In summary, a computer simulation is a useful tool in scientific research. Often it can provide information to help understand some processes. Numerical simulations, if done correctly, can be used to reduce the uncertainty/ambiguity and constraint of the interpretations. Sometimes, people are simply convinced by the beautiful movies made from simulations. This is a very dangerous mentality in science. I have personally benefited greatly by working with simulationists and developed various theoretical models to understand processes in space physics. However, I have to be very critical of every simulation result until I understand the processes it actually describes. I would be very suspicious if it is used as *direct proof* of a complex system. If the system is not very complex, it does not need numerical simulation to prove an idea, but rather to quantify it.

There are many non-first-principle numerical models, especially for economical or financial predictions; these cannot be used for scientific purposes or accounted as scientific.

Unacceptable proof: In space physics, there are two typical proofs that we do not accept.

The first is what we call "hand-waving". When a process is less than well understood, there is a large range of possibilities to explain an observation. Scientists have a large degree of freedom to propose arguments/assumptions (conjectures) without substantiation from the above three acceptable types, especially by evaluation. Hand-waving arguments, which often invoke some unspecified mechanisms, are useful to describe the proof but usually are not considered as a proof. To prove an idea is a much harder job than simple "hand-waving". One needs to substantiate the hand-waving with solid reasoning and evidence as well as evaluation.

The second is stand-alone neural network/artificial intelligence simulations. These efforts can describe or replicate some key parts of the statistics of an observation. It may even be useful to make predictions under normal conditions or identify some phenomena or correlations. But so far, they alone have not been able to provide an understanding of the problem. It may be better suited for engineering, business, or management problems but not for scientific proof, at the current time. Current neural network/artificial intelligence methods have to be able to reveal the mechanisms and understanding of the inner working of a complicated natural system in order to be tools for science.

11.4 Closing Remarks

When looking back at the conventional theories of philosophy of science, the fundamental flaw is the confusion about science, knowledge-understanding, and truth. Most theories assume science equals understanding which equals truth. They took science as knowledge and, hence, philosophy of science is a common problem of epistemology. Science is not about gathering or learning existing knowledge but about inventing new knowledge based on incomplete information. The focus is to make sure that the invented new knowledge is true. The present theories of epistemology do not provide guidance in this process, especially how to make judgments among multiple potential competing scientific theories. Most scientific results will not be able to reach the level of knowledge-understanding which is accepted by a large fraction of society. Therefore, existing theories of philosophy of science debate many issues irrelevant to science and scientists. Truth is objective whereas knowledge is a belief, i.e., subjective, and a condensed form of scientific results from multiple disciplines or subdisciplines. Because truth is objective and universal, but scientific results only reflect truth in some way, there is a fundamentally irreconcilable difference between the two. Most debates in philosophy of science have been about how to get around these inconsistencies based on concepts developed in epistemology.

At a level lower is the confusion among fact, theory, and evidence that, in general, involves "interpretation" between fact and theory. The uncertainty in the "interpretation" element results in the debate over the justification and falsification because neither can be conclusive when information is incomplete or insufficient. The "problem of induction" that has been debated about in philosophy of science is mostly caused by the confusion between generalization and prediction and between knowledge-that and knowledge-understanding. Knowledge-that resulting from simple generalization does not provide reliable predictions; knowledge-understanding has to be developed based on observations of different types, i.e., many different types of knowledge-that with underlying connections. Knowledge-understanding can substantially improve the reliability of a prediction although still cannot eliminate the possible false predictions because of the incomplete information available for the development of the understanding. Individual failures in prediction should not be used as evidence for questioning the process of human knowledge development. Instead, they can be used to identify the possible flaws or limitations of the science based on which the understanding was developed.

Science comprises three pillars: observation, conceptualization, and evaluation. The first two pillars provide information and inspiration. The last pillar is a requirement that distinguishes science from other methods of inquiry. It can substantially reduce the number of potential speculations. Many existing theories of philosophy of science have been based on "armchair speculations" and are irrelevant to science mostly because of the lack of evaluation control. Although to the general public, science is based on rationality and evidence so that it can correctly reflect the truth, to scientists, the problem is that there are multiple rationalizations and each piece of evidence is often based on a specific rationalization with an interpretation of a fact. Scientists need to make judgments among different possibilities of rationalization and interpretation with incomplete knowledge and partial information. Philosophy of science should provide a guideline for decision-making.

Science is driven by the invisible hand of curiosity of individual scientists and carried out with the natural selection process at the societal level. To scientists, philosophy of science is how to see the forest while studying trees, as Einstein stated. The most important questions of philosophy of science are: "How to get it right?" "How to know I am not wrong?" There is no scientific method. In science, there are two ways to get it right, experiment and conceptualization, and evaluation is used to control the process. Scientific reasoning is needed to connect the three. The two main forms of reasoning are scientific deductive and inductive reasoning. Scientific deductive reasoning is often used as the mainline of the reasoning, e.g., based on known natural laws and propositions. However, the limitations and conditions are essential for checking the applicability of the reasoning. Scientific inductive reasoning is a restricted form of inductive logic and often motivates a study; it is essential to determining the input to the deductive reasoning and from the output of the deductive reasoning to the conclusion. Any correlating phenomena have to be quantitatively defined. A simple yes–no correlation cannot guarantee the validity of a scientific conclusion. Although the "problem of induction" is real, the conclusion that induction cannot be used in science is fundamentally flawed.

More effort in science is made to address the question of "how to know I am not wrong" with scientific dialectical reasoning which is a restricted form of dialectical logic. The focus should be at the beginning of an approach, the interpretation of the result, and each point where the line of reasoning changes. One needs to investigate all "reasonable" and "relevant" possibilities and alternatives. When an alternative line of reasoning cannot be eliminated, it should be included in the conclusion. Occam's razor can be used to argue against the line of reasoning with more assumptions and changes in reasoning, as well as the one with more inductive reasoning or without clear evidence. Conventional theories of philosophers of science have made two major and unforgivable mistakes: to promote abductive reasoning as the main scientific logic — a fundamentally flawed logic in science — and to dismiss dialectical reasoning.

The common successful practice in science for other fields of inquiry to share or emulate is as follows: open debate based on scientific reasoning with quantitative control and taking reality as the final judgment for any ambiguity.

With these criticisms, recommendations, and new questions, I am expecting a new era of philosophy of science to come.

Philosophy of Science: Perspectives from Scientists

By Paul Song (UMASS Lowell)

Over the last hundred years, philosophy of science has developed its theory based on what philosophers perceived as science and how it is conducted. It does not address the basic questions that scientists care about. This book examines the conventional theories of philosophy of science from a completely different point of view and describes the most difficult problems that scientists are concerned about: How to get it right? And how to know I am not fooled by myself?

A theory of science is presented with examples that scientists encounter every day describing the process and organization of science and scientific reasoning with which scientists communicate in science. The book also provides in-depth theoretical discussions on the fundamental flaws in the conventional theories of philosophy of science which have produced misconceptions and confusion in society about science and scientists.

Bibliography

Abbot, F. E. (1885). *Scientific Theism*, University Press.

Alfvén, H. O. G. (1950). *Cosmical Electrodynamics*, International Series of Monographs on Physics, Oxford: Clarendon Press.

Bartlett, R. C. (2008). *Masters of Greek Thought: Plato, Socrates, and Aristotle*, The Great Courses, The Teaching Company, USA, www.TEACH12.com.

Bayes, T. (1763). An essay towards solving a problem in the doctrine of chances, *Philosophical Transactions of the Royal Society of London*, 53: 370–418. doi:10.1098/rstl.1763.0053.

Bickmore, B. (2010). Creativity in science, *Visionlearning.* POS-3(4). https://www.visionlearning.com/en/library/Process-of-Science/49/Creativity-in-Science/182.

Calendars through the Ages, http://www.webexhibits.org/calendars/year-text-Copernicus.html.

Carroll, J. B. (1997). Psychometrics, intelligence, and public perception. *Intelligence*, 24: 25–52. CiteSeerX 10.1.1.408.9146. doi:10.1016/s0160-2896(97)90012-x.

Copernicus, N. (1543). *On the Revolutions of the Heavenly Spheres,* translated with an introduction and notes by A. M. Duncan. London: David & Charles; New York: Barnes & Noble, 1976.

Cover, J. A., M. Curd, & C. Pincock, (1998). *Philosophy of Science: The Central Issues.*

Cowan, N. (2001). The magical number 4 in short-term memory: A reconsideration of mental storage capacity, *Behavioral and Brain Sciences,* 24(1): 87–114, discussion 114–185. doi:10.1017/S0140525X01003922.

Einstein, A. (1944). Albert Einstein to Robert A. Thornton, 7 December 1944, EA 61-574.

Encyclopedia Britannica (1989).

Feyerabend, P. K. (1975). *Against Method: Outline of an Anarchistic Theory of Knowledge*, Atlantic Highlands, NJ: Humanities Press.

Foster, R. N. (2015). What is creativity, https://insights.som.yale.edu/insights/what-is-creativity, Yale, February 16.

Galileo, G. (1967). *Dialogue Concerning the Two Chief World Systems*, translated by Stillman Drake, Berkeley: University of California Press.

Godfrey-Smith, P. (2003) *Theory and Reality: An Introduction to the Philosophy of Science*, Chicago: The University of Chicago Press.

Goodman, N. (1983). *Fact, Fiction, and Forecast*. Harvard University Press. Cambridge, MA. p. 74.

Gross, R. (2001). *Psychology: The Science of Mind and Behaviour*, UK: Hachette.

Gutting, G. (1980). Review of progress and its problems by Larry Laudan. *Erkenntnis*, 15(1): 91–103.

Hacking, I. (2012). Introduction by Ian Hacking, in *The Structure of Scientific Revolutions*, The London: University of Chicago Press.

Hempel, C. G. (1945). Studies in the logic of confirmation I, *Mind*, 54(213): 1–26. doi:10.1093/mind/LIV.213.1.

Hemper, C. G. (1966). *Philosophy of Natural Science*. Foundation of Philosophy Series, Englewood Cliffs: Prentice-Hall, Inc.

Hume, D. (1739). *A Treatise of Human Nature*, reprinted from the original edition in three volumes and edited, with an analytical index by L.A. Selby-Bigge (1986), Oxford: Clarendon Press.

Kant, I. (1781). *The Critique of Pure Reason.*, Translated by P. Guyer and A. W. Wood, Cambridge, Cambridge University Press.

Kasser, J. L. (2006). *Philosophy of Science*, The Great Courses, The Teaching Company, USA.

Kuhn, T. S. (1962). *The Structure of Scientific Revolutions*, London: The University of Chicago Press.

Lakatos, I. (1970). *History of Science and Its Rational Reconstructions*, In *PSA 1970*, Roger. C. Buck and R. S. Cohen (Eds.), Dordrecht: Reidel.

Latour B. & S. Woolgar (1979). *Laboratory Life: The Construction of Scientific Facts*, Princeton: Princeton University Press, (2nd Ed.), 1986.

Laudan, L. (1981). A confutation of convergent realism, *Philosophy of Science*, 48: 19–48.

Masterman, M. (1970). The nature of a paradigm, in *Criticism and the Growth of Knowledge*, Lakatos and Musgrave (eds.), *Proceedings of the 1965 International Colloquium in the Philosophy of Science*, vol. 4, Cambridge: Cambridge University Press, pp. 59–90.

Mendel, G. (1866). Versuche über Plflanzenhybriden. Verhandlungen des naturforschenden Vereines in Brünn, Bd. IV für das Jahr 1865, *Abhandlungen*, 3–47.

Merton, R. K. (1973). *The Sociology of Science: Theoretical and Empirical Investigations*, Chicago: The University of Chicago Press.

Miller, G. A. (1956). The magical number seven, plus or minus two: Some limits on our capacity for processing information, *Psychological Review*, 63(2): 81–97. CiteSeerX 10.1.1.308.8071. doi:10.1037/h0043158.

Moore, R. (2001). The "Rediscovery" of Mendel's Work, *Bioscene*, 27(2): 13–24.

NAS Committee on the Conduct of Science (1995). *On Being a Scientist: Responsible Conduct in Research*, Washington, DC: National Academy Press.

Nature Editorials (2007). Hard to swallow: Is it possible to gauge the true potential of traditional Chinese medicine? *Nature*, 448(July 12): 106.

Newton, I. (1687). *The Mathematical Principles of Natural Philosophy*, University of California Press, Berkeley, 1999.

Nola, R. & H. Sankey (2007). *Theories of Scientific Method: An Introduction,* Acumen Publishing, New York.

Okasha, S. (2016). *Philosophy of Science: A Very Short Introduction*, UK: Oxford University Press, Oxford.

Parker, E. N. (1996). The alternative paradigm for magnetospheric physics, *Journal of Geophysical Research*, 101(A5): 10587–10625.

Parker, E. N. (2007). *Conversations on Electric and Magnetic Fields in the Cosmos*, Princeton, NJ: Princeton University Press.

Philosophy of Science: The Central Issues. M. Curd, J. A. Cover, and C. Pincock eds., W.W.Norton and Company, New York, 1998.

Popper, K. (1972). *Conjectures and Refutations*, (4th Ed.). London: Routledge.

Popper, K. R. (2014). *The Logic of Science Discovery*, Mansfield, CT, USA: Martino Publishing.

Quine, W. V. O. (1951). Two Dogmas of Empiricism, *Philosophical Review*, 60: 20–43, reprinted in From a Logical Point of View, Cambridge, MA: Harvard University Press, 1953, pp. 20–46.

Quine, W. V. O. (1969). Epistemology naturalized, in *Ontological Relativity and Other Essays*. New York: Columbia University Press.

Robinson, D. N. (2004). *The Great Ideas of Philosophy*, (2nd Ed.). The Great Courses, The Teaching Company, USA.

Rosenberg, A., (2005). *Philosophy of Science: A Contemporary Introduction*, 2nd Ed., New York: Routledge.

Russell, B. (1918). *The Problems of Philosophy*, Oxford, Oxford University Press, 1959.

Simpson, E. H. (1951). The interpretation of interaction in contingency tables, *Journal of the Royal Statistical Society, Series B*, 13: 238–241.

Smolin, L. (2013). *There is No Scientific Method*, https://bigthink.com/articles/there-is-no-scientific-method/.

Song, P. & V. M. Vasyliunas (2014). Inductive-dynamic coupling of the ionosphere with the thermosphere and the magnetosphere. In *Modeling the Ionosphere-Thermosphere System*, J. Huba, R. W. Schunk, and R. W. Khazanov (Eds.), Geophys Monograph Series (Vol. 201, pp. 201–215). Washington, DC: American Geophysical Union.

Southwood, D. J. and M. G. Kivelson (1995). Magnetosheath flow near the subsolar magnetopause: Zwan-Wolf and Southwood-Kivelson theories reconciled, *Geophysical Research Letters*, 22(23), https://doi.org/10.1029/95GL03131.

Thurs, D. P. (2015). Myth 26: That the "scientific method" accurately reflects what scientists actually do. In *Newton's Apple and Other Myths about Science*.

Walktins, E. (2019). The laws of motion from Newton to Kant, *Kant on Laws*, Cambridge, Cambridge University Press. doi: https://doi.org/10.1017/978131668302.

Vasyliūnas, V. M. (2001). Electric field and plasma flow: What drives what? *Geophysical Research Letters*, 28(11), 2177–2180.

Index